Sustainability and Innovation

Coordinating Editor:

Jens Horbach

Series Editors:

Eberhard Feess
Jens Hemmelskamp
Joseph Huber
René Kemp
Marco Lehmann-Waffenschmidt
Arthur P.J. Mol
Fred Steward

Sustainability and Innovation

Published Volumes:

Jens Horbach (Ed.)
Indicator Systems for Sustainable Innovation
2005. ISBN 978-3-7908-1553-5

Bernd Wagner, Stefan Enzler (Eds.)
Material Flow Management
2006. ISBN 978-3-7908-1591-7

A. Ahrens, A. Braun, A.v. Gleich, K. Heitmann, L. Lißner
Hazardous Chemicals in Products and Processes
2006. ISBN 978-3-7908-1642-6

Ulrike Grote, Arnab K. Basu, Nancy H. Chau (Eds.)
New Frontiers in Enviromental and Social Labeling
2007. ISBN 978-3-7908-1755-3

Marco Lehmann-Waffenschmidt (Ed.)
Innovations Towards Sustainability
2007. ISBN 978-3-7908-1649-5

Tobias Wittmann
Agent-Based Models of Energy Investment Decisions
2008. ISBN 978-3-7908-2003-4

Rainer Walz • Joachim Schleich

The Economics of Climate Change Policies

Macroeconomic Effects, Structural Adjustments and Technological Change

Physica-Verlag
A Springer Company

PD Dr. Rainer Walz
Prof. Dr. Joachim Schleich

Fraunhofer Institute for Systems
and Innovations Research (ISI)
Breslauer Straße 48
76139 Karlsruhe
Germany
rainer.walz@isi.fraunhofer.de
joachim.schleich@isi.fraunhofer.de

ISBN 978-3-7908-2567-1 e-ISBN 978-3-7908-2078-2

Sustainability and Innovation ISSN 1860-1030

Cover design: WMXDesign GmbH, Heidelberg

Printed on acid-free paper

9 8 7 6 5 4 3 2 1

springer.com

Contents

1 Introduction 1

2 Effects of Climate Policy on the Economy: A Theoretical Perspective 5
2.1 Macroeconomic Objectives 5
2.2 Price and Cost Effects 6
2.2.1 Effects of Changes in Prices and Costs 6
2.2.2 Existence of No-Regret Potentials 8
2.2.3 Existence of a Double Dividend 13
2.3 Innovation Effects 15
2.3.1 Policy-Induced Technical Change 15
2.3.2 Productivity Effect of Investments in Climate Protection 16
2.3.3 First Mover Advantage 18
2.4 Demand Effects 25
2.4.1 Structural Changes 25
2.4.2 Income Multiplier and Accelerator Effects 28
2.5 Combined Effects of the Impulses 29

3 Empirical Results of the Macroeconomic Impacts 33
3.1 Modelling Approaches 33
3.2 Macroeconomic Effects 39
3.2.1 Review of the Results from Top-Down Approaches 39
3.2.2 Results Using a Combined Bottom-Up/Top-Down Approach 43
3.2.3 Comparison and Interpretation 47

4 Structural Adjustments 53
4.1 Scope of Analysis 53
4.2 Sectoral Changes 54
4.3 Changes in Regional Employment 61
4.3.1 Scope of Analysis and Methodological Approach 61

4.3.2 Net Changes in Employment in the Labour Office District ... 62
4.3.3 Gross Changes in Number of Jobs ... 64
4.3.4 Comparison of East and West Germany ... 65
4.3.5 Effects on the Regional Concentration of Employment ... 67
4.4 Changes in Qualification and Working Conditions ... 67

5 Theoretical Approaches and the State-of-the-Art in Accounting for Technological Change ... 71
5.1 Definitions ... 71
5.2 Neoclassical Innovation and Environmental Economics ... 73
5.3 Evolutionary and Institutional Economics ... 75
5.4 Systems of Innovation Approach ... 77
5.5 Regulatory Economics ... 79
5.6 Policy Analysis ... 80
5.7 Review of Empirical Results ... 81

6 Technological Change and Energy Consumption in Energy Intensive Industry ... 87
6.1 Technological Change and Energy Consumption in the German Manufacturing Sector ... 87
6.1.1 Energy Consumption of the German Manufacturing Sector ... 87
6.1.2 Decomposition Analysis for Energy Consumption and Energy Intensity in the Manufacturing Sector ... 90
6.1.3 Estimations of Determinants for the Development of Fuel Intensity ... 92
6.2 Econometric Analyses for the Iron and Steel Sector ... 95
6.2.1 Technological Paradigms ... 95
6.2.2 Estimations of Aggregate Fuel Intensity in the Production of Pig Iron ... 98
6.2.3 Estimations for the Diffusion of the Electric Arc Furnace ... 99
6.2.4 Estimations for the Development of Energy Efficiency ... 101
6.3 Conclusions ... 106

7 Barriers to the Diffusion of Energy Efficiency in the German Commercial and Services Sectors ... 107
7.1 The Data ... 108
7.2 Barriers to the Diffusion of Energy Efficiency ... 109
7.2.1 Lack of Time ... 110
7.2.2 Lack of Information About Energy Consumption Patterns ... 111
7.2.3 Lack of Information About Measures ... 111

7.2.4 Investment Priorities ... 111
7.2.5 Uncertainty About Future Energy Costs ... 112
7.2.6 Landlord/Tenant Dilemma (Split Incentives) ... 112
7.3 Determinants ... 113
7.3.1 Energy Consumption ... 113
7.3.2 Size ... 113
7.3.3 Energy Audits ... 113
7.3.4 Sub-Sector Dummies ... 114
7.4 Results ... 115
7.5 Conclusions ... 117

8 Innovation Effects of Regulation – Case Study for Wind Energy ... 119
8.1 Introduction and Methodology ... 119
8.2 Regulation in the Wind Energy Sector ... 122
8.2.1 Overview of Instruments ... 122
8.2.2 Regulation in Germany ... 124
8.2.3 Regulation in the US ... 126
8.3 Development and Diffusion of Wind Turbines ... 128
8.3.1 Development of Technology ... 128
8.3.2 Diffusion of Technology ... 131
8.4 Comparative Analysis of the Influence of Regulation on Innovation ... 133
8.4.1 Experimentation Phase in the 1980s ... 133
8.4.2 Rapid Market Growth in the 1990s ... 134
8.4.3 Future Outlook ... 138
8.5 Questions for Further Research ... 140

9 Summary and Conclusions ... 143

Appendix: Model Description of ISIS ... 153

Literature ... 155

Chapter 1
Introduction

In its latest Assessment Report, the Intergovernmental Panel on Climate Change (IPCC, 2007) projects that without further action the global average surface temperature would rise by a further 1.8–4.0°C until the end of this century. But even if the rise in temperature could be limited to the lower end of this range, irreversible and possibly catastrophic changes are likely to occur. Consequently, the protection of the earth's atmosphere requires substantial efforts to reduce CO_2 and other greenhouse gas emissions – especially in countries with very high per capita emissions. To limit the imminent rise in temperature, in the Kyoto-Protocol, the European Union has committed itself to reducing the emissions of greenhouse gases by 8% up to 2008–2012 compared to 1990 levels. Within the EU burden sharing agreement, some countries have to achieve even higher emissions reductions. Germany was assigned a reduction target of 21%. The entry into force of the Kyoto Protocol in February 2005 marks a first step towards meting global climate targets, but more ambitious action to reduce greenhouse gas emissions is needed after 2012, when the Kyoto targets expire. Under German presidency, the EU has committed itself to unilaterally reduce its greenhouse gas emissions until 2020 by 20%. In case a Post-Kyoto agreement can be reached, the EU reduction target would be 30% (CEU, 2007). Such reduction targets would be on a path towards meeting the 70–80% emission reductions considered necessary by 2050 for industrialised countries to meet long term global climate targets, taking into account economic growth in many populous developing countries. To meet these targets governments have started to implement climate policies which include economic instruments such as energy or emission taxes, or CO_2-emission trading schemes, subsidies for the invention, adoption and diffusion of low- or no-carbon technologies such as renewable energy sources, efficiency standards or labelling systems for appliances, or informational measures such as demonstration programmes. These policies change the incentives structure of economic agents towards more climate-friendly processes and products. They also alter the profitability of new climate-friendly technologies, leading to additional research and development efforts towards such technologies, eventually resulting in a less carbon-intensive production and consumption structure of the entire economy.

In particular, the anticipated economic effects of these policies play an important role in shaping the political debate over climate protection policies, which has been

R. Walz and J. Schleich. *The Economics of Climate Change Policies.*
Sustainability and Innovation,

refuelled by the findings of the "Stern" report in 2006 (Stern, 2007). Accordingly, without further action, the overall costs of climate change will be equivalent to losing at least 5% of global gross domestic product each year, while the costs of reducing greenhouse gas emissions to avoid the worst impacts of climate change could be limited to about 1% of global GDP per year. However, these findings as well as the implications for climate policy are highly disputed (Tol, 2006; Nordhaus, 2007; Heal, 2008). Similarly, recent efforts to assess the macroeconomic implications of climate policies, for example on employment or gross domestic product, have not produced a clear picture. Conflicting model results have contributed to the evolvement of an intensive debate on the costs of climate protection. "The Economics of Climate Change Policies: Macroeconomic Effects, Structural Adjustments and Technological Change" portrays this debate, analyses the reasons behind different modelling results, and highlights the weaknesses of existing studies. Furthermore, it presents its own empirical results which contribute to closing the gap with regard to structural effects and offers new insights into the modelling of technological change.

Chapter 2 of the book analyses the economic mechanisms which are responsible for the macroeconomic effects of climate protection policies. Three different classes of mechanisms with various subgroups can be distinguished: effects on costs (supply side), effects on aggregate demand, and technological effects (productivity, technological competitiveness). A sound theoretical explanation is given for why different results can emerge, depending on which mechanisms are taken into account. This theoretical chapter concludes with a hypothesis about the likely pattern of economic impacts of climate protection policies.

Based on this framework, the different approaches and results of modelling economy-energy interactions are evaluated in Chap. 3 for the case of Germany. It is demonstrated that different modelling approaches emphasise different segments of the numerous economic mechanisms. The most important empirical studies in Germany are analysed, and differences between the results are explained. Based on this analysis, conclusions are drawn about the likely macroeconomic impacts of climate change policies.

Chapter 4 deals with the structural adjustments of climate change policies. The pattern of sectoral changes in the industry structure are analysed for two policy scenarios which differ with regard to the policy instruments assumed. Furthermore, both the effects on changes in the qualitative job characteristics and qualification requirements as well as the on regional adjustments in employment are analysed.

Taking into account innovation effects is one of the key weaknesses of modelling economic effects. Thus, Chap. 5 addresses the issue of technological change in more detail. Various approaches to explain the generation and diffusion of new technological solutions are compared, and the state-of-the-art of existing empirical studies is summarised. In the following three chapters, case studies are performed to explore the empirical relevance of the various theoretical approaches for the determinants of technological change in the context of energy use. Chapter 6 includes two empirical case studies form the industry sector in Germany using econometric techniques to explore the determinants of technological change in the production of energy-intensive products. The first case study focuses on the impact of energy prices on energy use in the German manufacturing sector. The second case study analyses the determinants for the development and the diffusion of new technologies in one of the main energy-consuming industrial sectors, the steel industry. More detailed analyses are conducted in order to further assess the impact of various determinants of intra-sectoral structural change, and of the adoption and diffusion of more energy-efficient technologies for the main technological paradigms, basic oxygen steel production and electric arc furnace steel. Besides energy prices the determinants considered also include expenditure for research and development, industry concentration or sunk costs. In Chap. 7, a third case study deals with the impact of obstacles to a rational energy use, and the role of soft measures such as energy audits in sectors where energy costs play a minor role only, since the processes to produce goods and services are less energy intensive. More specifically, econometric methods are applied to empirically assess the relevance of various factors of influence on barriers to the diffusion of energy efficient technologies and measures in the German commercial and services sectors. The barriers considered include lack of time, lack of information, uncertainty about energy costs and split incentives for investments in energy efficiency. In particular, the relevance of "hard" factors such as energy costs or company size is assessed together with the impact of the rather "soft" measure energy audit. In Chap. 8, the final case study broadens the policy perspective to the interplay of regulation and innovation within a wider system of innovation approach. The fourth case studies deals with innovations in the wind sector. It addresses the question how different forms of policies and regulations might influence the generation and diffusion of new technical solutions. In contrast to the previous case studies, it does not rely on econometric techniques, but refers to evidence obtained by qualitative research such as interviews and questionnaires combined with – theory based – interpretation. Each of the case study chapters includes a brief summary and stands by itself.

The final chapter (9) summarises and combines the main results and insights of the previous chapters to assess the empirical relevance of the different theoretical approaches to technological change, provides policy recommendations and offers suggestion of how to further improve economic models to adequately capture innovations.

Chapter 2
Effects of Climate Policy on the Economy: A Theoretical Perspective

2.1 Macroeconomic Objectives

The aspects which are relevant when assessing the economic and social compatibility are almost unlimited. Theoretical studies look at changes in the macroeconomic welfare. The topics examined in economic-policy analyses range from macroeconomic variables, regulatory frame conditions, income distribution, and structural change through to the conditions which can be derived with regard to social compatibility.

In order to reduce the analyses to be carried out to a manageable amount, it is necessary to concentrate on those problem areas where it is supposed that climate policy could have significant effects. Since climate policy involves sectorally overlapping measures with which considerable investment sums are induced, the impacts on the *macroeconomic variables* play an important role. Here, the impacts on the national or domestic product are often interpreted as a reference value for the macroeconomic welfare, since a continuous growth of the national product implies that the sum of the goods available for consumption is continuously increasing. Thus, the effects on the Gross Domestic Product (GDP) will be used to assess the macroeconomic impacts. However, in doing so it is important to keep the limitation of this indicator in mind. The national product – even without referring to the external costs of climate change[1] – is only a very indirect yardstick for measuring welfare. The following aspects have to be considered:

- The GDP measure does not express which share is allotted to the supply of commodities to the population. It is possible, for example, that consumption is reduced in spite of a constant national product, if investments increase as a result of climate protection.[2]

[1] When assessing the impacts of climate policy on the macroeconomic welfare, the avoided impacts of climate change also have to be taken into account. The assessment of policy consequences, however, concerns quantifying the macroeconomic effects of climate protection in order to be able to compare these with the benefits of climate protection policy. When the term "welfare" is used in this section, therefore, this does not include the external costs of climate change.

[2] From a theoretical point, any assessment of the welfare changes of a CO_2 reduction based on tangible goods supply should therefore also take into account the changes in aggregated private consumption.

R. Walz and J. Schleich. *The Economics of Climate Change Policies.*
Sustainability and Innovation,

- The GDP measure does not reflect either a change in the commodity leisure or in non-market evaluated activities.
- Changes in GDP do not capture changes in the external costs. A climate policy does not only reduce greenhouse gas emissions, but also brings about a reduction of other external costs (e.g. emissions of air pollution). Likewise, to the extent that climate policy results in lower imports of fossil fuel from politically sensitive regions, climate policy also contributes to increase security of supply. These kinds of secondary benefits are neither included in the national product nor in the benefits of reducing the CO_2 emissions.[3]

The impacts on *employment level* constitute an essential topic when assessing social compatibility, especially in times of high unemployment. This variable is therefore subject to closer scrutiny, and is used as the second major indicator to assess the macroeconomic effects of climate policies.[4]

This section seeks to answer the question: due to which economic mechanism can the implementation of climate policy measures result in changes of the GDP and the number of jobs? Climate policy measures set off diverse adjustment reactions among individual companies and private households which precipitate as structural effects on a sectoral and regional level. The sum of these adjustment reactions and the subsequently caused impacts then result in changes of macroeconomic variables on the macroeconomic level. The various economic mechanisms describe which adjustment reactions and consequential effects are induced by climate policies. However, they are strongly influenced by the respective theoretical paradigm used. In line with the various schools of thought, price and cost effects, demand effects and innovation effects can be distinguished.

2.2 Price and Cost Effects

2.2.1 Effects of Changes in Prices and Costs

Price and cost effects stand at the forefront of neoclassical economic theory. The primary cost factors in the general economic discussion are the costs of labour (wages) and capital. With regard to climate policy, another cost factor is the cost occurring for supplying energy services. It is true that, in a macroeconomic context,

[3] Accordingly, such effects would also have to be taken into account in an overall assessment of climate policy.

[4] Furthermore, sectoral or regional structural changes cause adjustment pressure in both growing and shrinking sectors and are therefore of considerable significance for both the social compatibility and when assessing the political enforceability of a climate policy. However, they are not included in Chap. 3 on macroeconomic effects, but are dealt with in Chap. 4.

these only comprise a small share of total costs, but they are strongly affected by changes in the energy supply or by climate protection.

If restrictions are placed on CO_2 emissions, the use of fossil fuels has to be reduced. If the cost burden increases as a result of this, various supply side effects will be triggered.[5] The same amount of labour is demanded for a lower real wage. Furthermore, substitutions in favour of other production factors take place, and, in general, the potential output is reduced. If the market mechanism on the labour market leads to lower real wages, a new equilibrium with full employment is reached. If this is not the case, unemployment will result. Furthermore, an increase in costs leads to disadvantages in international competition.[6] If, on the other hand, the climate policy brought about cost reductions, an increase of production and employment would result and international competitiveness would be improved. From these arguments it is clear that, within the scope of neoclassical theory, it is decisive for the direction these cost effects take whether a climate protection policy results in an increase or a reduction of the cost burden.

Neoclassical theory assumes, in general, that market mechanisms lead to an efficient allocation of resources. If such an efficient starting-point is assumed, a reduction of CO_2 emissions, in general, leads to a reduction of GDP.[7] In the scope of this static analysis, a different result can only come about if the assumption of an efficient starting-point is abandoned. The discussion of the effects of climate protection policy is therefore distinguished by the – often implied – characterisation of the starting situation. A *non-efficient starting situation* is accounted for on two different aggregation levels:

- The so-called no-regret potential is picked out as a central topic on a technology-based aggregation level.[8] It is argued that a considerable cost-efficient potential to reduce greenhouse gas emissions already exists under the given frame conditions. From this it is concluded that, to a certain extent, a reduction in greenhouse gas emissions would be possible without increasing the cost burden.
- Specifically with regard to the introduction of an energy or CO_2-tax, it is argued that a double dividend (emissions reduction plus increase in economic output) could be achieved if the tax revenue is used to lower other distorting taxes.

[5] See Lintz (1992, pp. 34–38) and Landmann (1984, pp. 181–190) based on the example of an increase in oil price.

[6] This effect is reduced if there are internationally agreed reductions such as those aimed at in the Kyoto Protocol. On the other hand, a first-mover advantage may result from unilateral (national) acts. See Sect. 2.3.3.

[7] In theoretical terms, an efficient starting-point is equivalent to a position on the production-frontier of an economy. Any increase in the output of one composite of the total output inevitably requires the reduction of at least one other composite, reducing welfare. If one neglects the problems that GDP is not able to account for all welfare changes (see Sect. 2.1), this is equivalent to a reduction in GDP.

[8] For a comprehensive definition of the no-regret potential see Ostertag (2003).

2.2.2 Existence of No-Regret Potentials

The existence of no-regret potentials is supported by comprehensive technology-based analyses in which cost-efficient saving possibilities can already be shown for numerous case examples under the given frame conditions.[9] Information deficiencies and investor's demand for very short term payback times are given as the main reasons for the existence of these untapped saving potentials.[10] Other obstacles cited include principal agent problems within companies as well as – especially relevant for housing construction – the non-appropriability of profits, which is caused by asymmetric information.[11] In addition to this, for the electricity sector, reference is made to the former kind of regulation of natural monopolies. They used to restrict measures of electricity conservation through degressive tariff designs on the one hand. On the other hand, electric utilities used their market power to frighten away environmentally-friendly independent electricity producers from entering the market.[12]

The concept of no-regret potentials was considerably extended and expanded to environmental protection in general by the work of Michael Porter.[13] He argues that a strict environmental protection policy can result in companies activating unused efficiency potentials and thus achieving competitive advantages (*free lunch hypothesis*). This theory is embedded in a concept of the international competitiveness of nations, which declares that the main factors of determination are found in the interaction of production factor availability, business strategies, demand conditions, industry clusters and supporting government action leading to changing frame conditions.[14] Intensive competition and the constant search for more efficient solutions are essential to achieve competitive advantages. In this sense, it can even be advantageous for a country if it starts off at a competitive disadvantage due to missing production factors, since this encourages innovations and avoids the squandering of resources. Increasing the stringency of environmental standards can work in a similar way: "The notion is that the imposition of regulations impels firms to reconsider their processes, and hence to discover innovative approaches to reduce pollution and decrease costs or increase output."[15]

---

[9] See Grubb et al. (1993, pp. 403–432), IPCC (1995, pp. 309–312, 317–322), IPCC (2001, pp. 504–512).

[10] See Jaffe and Stavins (1994b, pp. 804–806), Sanstad and Howarth (1994, pp. 814–816), Metcalf (1994, pp. 821–823), Cameron et al. (1999, pp. 66–69). For a recent overview and conceptual framework on those barriers to energy efficiency, see Sorrell et al. (2004).

[11] DeCanio (1993, pp. 907–910), Cameron et al. (1999, p. 65). In the literature, this effect is known as the "investor-user dilemma". An investment by the owner in energy conservation cannot be transferred into higher basic rents, since new tenants can only estimate the energetic condition of a building at a high information cost.

[12] See e.g. Brunekreeft (2004).

[13] Porter (1990); Porter and van der Linde (1995).

[14] Berg and Holtbrügge (1997, p. 200).

[15] Jaffe et al. (1995, p. 155).

Both the results of bottom-up case studies, the associated explanations and Porter's reasoning are challenged theoretically. *"There is no free lunch"* is the counter-argument to the existence of a no-regret potential.[16] Basically, the arguments of the no-regret opponents whittle down to one or another form of hidden costs, which the proponents "forget" to account for.

In turn there is a whole series of counter arguments to this line of thought. The main ones are those which put forward a different theoretical starting-point. For example, in certain cases, the existence of no-regret potentials can be justified from the perspective of transactions cost economics and real options value theory.[17] In addition, numerous publications question the assumption of the utility maximising, rational behaviour of people and are emphasising – in addition to information and knowledge deficiencies – the competence and motivations of actors and putting forward the concept of *bounded rationality* instead.[18] In addition, from the perspective of evolutionary economics, the influences on the decision process at different levels have been highlighted and the assumption of fixed and transitive preferences has been challenged.[19] Finally, it is pointed out that decisions within companies are the result of a complex process, which is characterised by multifunctional network structures with differing objective functions, spillovers between the individual sectors and limited information processing abilities so that, at any time, there is the possibility to bring about substantial efficiency improvements.[20]

Thus, there are numerous arguments in the debate on the existence of no-regret potentials which are expressions of differing scientific paradigms. However, the existence or non-existence of no-regret potentials cannot be as clearly proven as the representatives of the respective positions would like to claim. Overall, the conclusion drawn from these models is that there are no strict behavioural assumptions: "The evidence and models surveyed suggest that a sensible rationality assumption will vary by context, depending on such conditions as deliberation cost, complexity, incentives, experience, and market discipline."[21]

For the assessment of climate protection policy, the consequence is that the existence of inefficiencies is a necessary but not sufficient condition for empirically significant no-regret potentials of a climate policy. In order for a climate protection policy to systematically reduce costs in proportion to other conceivable policy fields, the inefficiencies in energy consumption have to be particularly pronounced. In addition, there must be policies available to systematically and cost-effectively remove these inefficiencies. Thus, arguments are necessary which support the supposition

---

[16] See Palmer et al. (1995, p. 120).

[17] See Ostertag (2003).

[18] See Conslik (1996, pp. 669–683). In Simon (1997, p. 291), bounded rationality refers to "cognitive limitations of the decision makers".

[19] See Witt (1987, pp. 133–137), Vanberg (2001).

[20] Nelson (1995, pp. 51 ff). This argument is applied to energy and climate policy in the work of Dennis et al. (1990), Stern (1992), DeCanio and Watkins (1998a, b) as well as DeCanio et al. (2000, 2001).

[21] Conslik (1996, p. 692).

that inefficiencies exist, particularly with regard to decisions about energy saving investments. Alongside the already mentioned traditional reasons for market failure, a justification may exist in a form of bounded rationality which does not adapt fast enough to the changed frame conditions and therefore forfeits its efficiency, which may well have been present under the original conditions. The following aspects must be considered here:

- The company's energy supply is not at the centre of the corporate performance processes. In the sense of satisfying, the aspiration level consists mainly in securing supply at reasonable costs.
- During times of sinking energy costs, routines developed stating that a costly search for energy saving possibilities does not pay off anyway. This decision routine is plausible for the large number of companies in which energy consumption mainly occurs in ancillary services such as the supply of process heat or compressed air production and only constitutes a small share of the total costs.[22]
- This tendency is reinforced even more by the fact that energy-relevant investment decisions are often reinvestments, in which the decision is not made independently of decisions taken at earlier points in time. Thus, inefficient decisions in the past influence future decisions (good money is thrown after bad).
- Energy-relevant investments do not have to be made continuously, but often have a rather ad hoc nature. At the same time, these investments also often have a disproportionately long lifespan. On top of this are the complex environment and uncertainties with regard to future developments. Under these conditions it is especially plausible to have an orientation along decision routines which are difficult to dismantle, even more so since the drop in energy prices after the oil price crises seems to reinforce these kinds of decision routines.
- Policy measures which draw attention to the necessity to reduce CO_2 emissions in the future could help to change these decision routines independently of whether they alter relative prices or have a different leverage.[23] The altered decision routines establish themselves through social interaction. For measures to accelerate the diffusion of these routines, it is of considerable significance to select the respective multipliers and opinion leaders as the target group. This type of group- specific design can simultaneously limit the costs of the policy.

To sum up, the arguments presented above may be sufficient to explain why a certain no-regret potential exists. Such a no-regret potential also influences the shapes of the *cost curves* of a CO_2 reduction. If one assumes an efficient economy, every option with costs below zero will be realised. Thus, Point T in Fig. 2.1 will be realised. It characterises the theoretical cost minimum an efficient economy will reach.[24] If an additional restriction is added, such as a reduction of CO_2

[22] This argument is consistent with empirical results that relatively high unused saving potentials are found mainly in companies with low specific energy consumption. See Chap. 5.2 of this book.

[23] DeCanio (1999, pp. 291–292).

[24] As mentioned, external effects are excluded here.

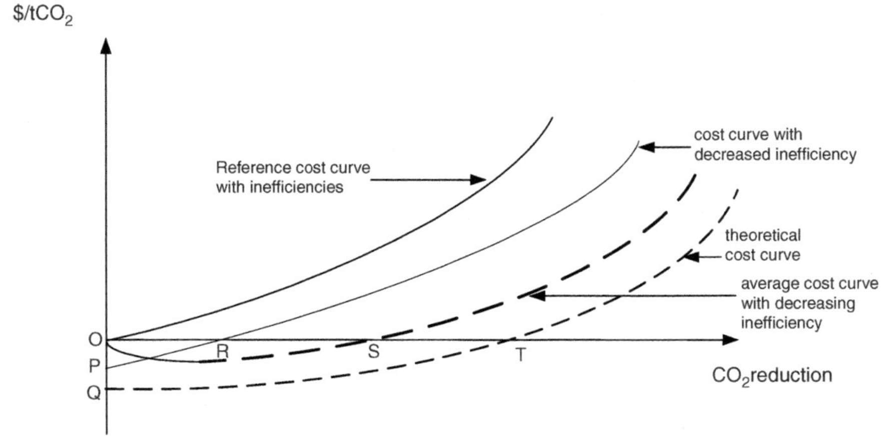

Fig. 2.1 Costs curves of CO_2 reduction including inefficiencies

emissions, this leads to an increase in costs. Thus, the theoretical (marginal) cost curve of CO_2 reduction has a positive slope to the right of starting point T.[25]

However, if it is assumed that inefficiencies exist in the reference case, there is another starting point. In contrast to an efficient economy, the emission reduction characterised by OT is not realised due to inefficiencies.[26] In an inefficient economy, the cost curve for the reference case starts in O, and the inefficiency is characterised by the area OQT. Thus, if one compares an inefficient economy with a perfect efficient economy, the theoretical no-regret-potential is equal to the emission reduction OT.

The arguments for the existence of a no-regret potential all relate to the inevitability of a certain amount of inefficiencies. Thus, the starting point for the analysis is point O on the reference (marginal) cost curve with inefficiencies. However, if it is possible to reduce the inefficiencies during the course of the climate policy, for example because decision routines are changing, the cost curve for the policy case is modified and shifts to the right towards the theoretical cost minimum curve. Nevertheless, it seems unrealistic that the theoretical cost minimum curve can be reached, not least because the measures to increase efficiency still incur costs, even in the most favourable case, or because the inefficiency is not completely removed. In Fig. 2.1, it is assumed that the inefficiencies are at least partially reduced. This corresponds to a (marginal) cost curve with decreased inefficiencies starting at point P.

However, the utilisation of a no-regret potential of climate policy is characterised by both a reduction of inefficiencies (leading to a shift of the marginal cost

[25] It can be assumed that increasing reductions of CO_2 will be associated with increasing marginal cost. Thus, the positive slope of the reduction curve is increasing.

[26] From the viewpoint of an efficient starting-situation, the CO_2 reduction costs are negative up to point T and the associated reduction is carried out. Correspondingly, the section of the cost curve relevant for an efficient starting situation does not begin until point T.

curve), and utilisation of technologies with increasing marginal costs (movement along a marginal cost curve). Combining both effects in one cost curve yields an average cost curve with decreasing inefficiency.[27] Starting from point O, there is a reduction in average costs until point R is reached, because this emission reduction is associated with negative marginal costs. The maximum efficiency gain is realised at point R (intersection of marginal cost curve with decreased inefficiency with the axis) and equals the area OPR below the marginal cost curve with decreased inefficiency. At the same time, the average costs curve reaches its minimum, if the emission reduction R is realised. However, the average costs are still below zero if emissions continue to be reduced. At point S, however, the reduction of average costs has been used up by the increase in marginal costs of the emission reduction beyond point R. The emission reduction OS realised in comparison to the reference case is the "no-regret" potential. It is defined as the emission reduction which can be realised without additional total costs compared to the reference case. However, in contrast to the situation characterised by OR, there is no reduction in total costs. Instead the increase in efficiency is used to realise further emission reductions.

With stricter environmental targets, however, measures at much higher costs have to be taken. The higher the targeted CO_2 reduction, the stronger the influence of the rising marginal costs of CO_2 reduction on the development of the average cost curve and the weaker the cost-reducing effect of decreasing the inefficiencies.

For the assessment of the economic impacts of climate policy, it is decisive at what amount the no-regret potential is estimated. Based on the estimations for North America and Europe up to the mid-90s, which were made primarily using technology-based energy system models, a no-regret potential equalling 10–30% of the CO_2 emissions was identified.[28]

When interpreting these results, it must be kept in mind that they generally refer to the maximum no-regret potential achievable under the assumed technological change,[29] i.e. to the range OT in Fig. 2.1. With regard to estimating the economic compatibility, it would however be necessary to estimate the potential OS, which is easily realisable through policy measures.[30] As long as there are no empirically sound, comprehensive estimates available, the no-regret debate ultimately remains

[27] In order to keep the argument simple, the explanation of Fig. 2.1 only assumes a singular decrease in inefficiency. If, however, the decrease in inefficiency happens continuously, the slope of the average cost curve decreases and it shifts towards the theoretical cost curve.

[28] Grubb et al. (1993, p. 470); the final conclusion of the IPCC is cited all over the world: IPCC (1995, p. 12): "Despite significant differences in views, there is an agreement that energy efficiency gains of perhaps 10 to 30% above baseline trends over the next two to three decades can be realised at negative to zero net cost." See Koomey et al. (1998), Loulou et al. (2000), Krause et al. (1999), Brown et al. (2000), IPCC (2001) for more recent statements.

[29] However, these estimates tend to underestimate the future technological change and are thus comparatively too pessimistic. See Jochem (1997b), Seebregts et al. (2000), IPCC (2001, p. 512).

[30] Here it would be necessary to substantially extend the prior energy systems models which were customary up to now to cover transaction costs and hidden costs and benefits. See Jochem and Diekmann (2001).

stuck in a "lasting controversy between believers and non-believers in the existence of a large untapped efficiency potential in the economy".[31]

2.2.3 Existence of a Double Dividend

One component of strategies to increase energy efficiency is usually an increase in energy prices through *an energy/CO_2 tax*, which is compensated for by lowering other taxes; i.e. is conceived as being revenue-neutral. Since the relative prices of the production factors are altered by the levying of almost any tax, excess burden of taxation occurs.[32] Partial analysis shows that this excess burden can be diminished if a tax is replaced by an energy tax and if the energy tax has a lower excess burden than the existing tax, i.e. it is less distorting. If this is the case, the result is a positive revenue recycling effect leading to a double-dividend of tax reform.[33] The taxes on labour are judged to have strong distortional effects. Thus, the chances are quite high that an ecological tax reform, which reduces the taxes on labour, will yield a double dividend. The reduction of the excess burden of taxation diminishes (weak double dividend) or even overcompensates (strong double dividend) the direct costs of CO_2 reduction.[34]

However, in more recent publications, the issue is raised that the probability of a double dividend is very strongly influenced by the interaction of climate policy with a pre-existing tax system characterised by distortions.[35] General equilibrium analysis holds that a reduction of CO_2 emissions leads to lower production and reductions in employment or real wages. Keeping total tax revenue constant requires an increase in tax rates. If these taxes have a distortionary effect, the overall excess burden increases still further. This effect is called the tax interaction effect. It is still possible that an ecological tax reform might induce a revenue recycling effect, if the eco-tax revenue is used to lower other distortionary taxes.[36]

[31] IPCC (2001, p. 506).

[32] The excess burden of a tax is the loss in welfare which occurs, because almost any tax leads to distorting changes in relative prices which induce substitution effects.

[33] See Schöb (1995). The first dividend consists of the reduction of the environmental burden, the second of the positive economic impacts.

[34] Deviating from this definition, which is based on Parry et al. (1999) and IPCC (2001, p. 472), sometimes an increase of employment is characterised as a weak second dividend in the literature, and an increase of the GDP as a strong second dividend.

[35] See Bovenberg and de Mooij (1994), Bovenberg and van der Ploeg (1994), Goulder (1995), Parry et al. (1999).

[36] Thus, it can be argued that climate policy instruments which lead to revenues, which are used to lower distortionary taxes, have more favourable economic effects than instruments without revenues (see Parry et al., 1999). According to this argument, for example, energy/CO_2 taxes or emissions trading systems with auctioned allowances would have to be favoured against an emissions trading system with a grandfathering scheme.

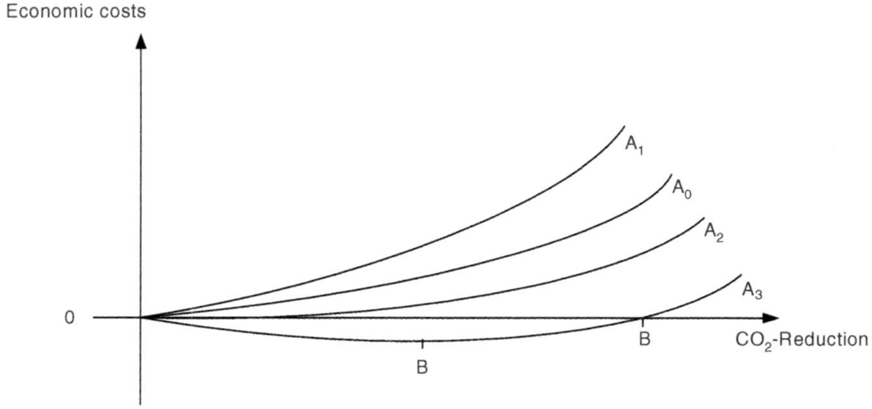

A_0: Direct (technology-related) costs without CO_2-Tax
A_1: Net costs under a CO_2-tax when tax interaction-effect is dominating
A_2: Net costs under a CO_2-tax when distortionary taxes are reduced (weak double dividend)
A_3: Net costs under a CO_2-tax when gross costs are partially overcompensated (strong double dividend)

Fig. 2.2 Conceivable impacts of a CO_2/energy tax on the national economy (adapted from IPCC 2001, p. 513)

However, in order to reach a strong double dividend, the revenue recycling effect must outweigh the tax interaction effect.[37]

In several theoretical analyses, it is deduced that the tax interaction effect dominates the revenue recycling effect.[38] However, this result is not universally valid, but depends on the assumptions made, i.e. the role of labour as only source of income, or the magnitude of cross price elasticities of substitutes to the produced goods. It is certainly possible that, based on different assumptions, situations can be modelled in which a strong double dividend may still occur.[39]

Overall, it can be concluded that the existence of a double dividend is controversial and does not represent a definite result. Figure 2.2 makes this clear: depending on the assumptions selected, the theoretical modelling can result in the direct costs of climate protection shown in curve A_0 being increased, e.g. if tax interaction dominates (curve A_1). On the other hand, the direct costs can also be softened by a weak double dividend (curve A_2) or even partially overcompensated by a strong one (curve A_3).

[37] IPCC (2001, pp. 472–473). These arguments are ultimately based on a second-best problem. Since several optimality conditions are violated, the removal of one inefficiency (revenue recycling effect) does not at the same time necessarily result in an improved allocation.

[38] See Bovenberg and de Mooij (1994), Parry et al. (1999).

[39] See Koskela et al. (2001, pp. 21–29), Parry and Bento (2000) and Bosello et al. (2001, pp. 15–17). Correspondingly, the argument is also put forward that the conditions for the existence of a double dividend are comparatively more favourable in many European countries than in the USA, IPCC (2001, p. 516). Furthermore, recent analyses argue that, under certain conditions, a double dividend may arise due to shifting the burden to international trade partners; see Smulders (2001).

These comments on ecological tax reforms refer to the impacts on the GDP. For effects on employment, the *substitution effects* have to be taken into account as well, which result from an energy/CO_2 tax financing the reduction of the taxes on labour – e.g. social security contributions. The production factor labour becomes cheaper in relation to the other production factors as a result of this reduction. Because of this change in the relative prices, an incentive then exists to employ more labour and substitute other production factors in this way. Whether employment increases under depends on whether these substitution effects are large enough to overcompensate the decreases in employment resulting from a drop in total production. It can be stressed that the impacts of an ecological tax reform on employment tend to be more positive than those on the GDP.

2.3 Innovation Effects

2.3.1 Policy-Induced Technical Change

Up until now, most analyses of the costs of climate protection assumed that technological change takes place without any impacts from the climate protection policy and that an autonomous increase in the energy efficiency of the national economy results over time.[40] If, however, due to the climate policy, additional technological change is induced, given emission reduction targets can be reached in a more cost-effective way.[41]

To what extent a climate policy can actually influence *the generation of new innovations* is of significance when assessing this effect. From the perspective of environmental economics, a corresponding innovation effect is assigned to market economic instruments.[42] Consequently, it can be reckoned that a climate policy which increases energy prices will result in corresponding effects. However it remains unclear how regulatory instruments effect the generation of innovations. They play an important role, for example, in the space heating sector, in which a considerable share of CO_2 reductions is to be achieved. Against the background of the innovation-hindering effect which is generally assigned to standard-based instruments by environmental economics,[43] the question arises of whether it is likely that any incentives to generate new technical solutions will occur. Answering this question is of considerable significance for the entire discussion of the costs of

[40] This corresponds to a shift of the production frontier, in which the marginal rate of transformation would be altered in a way that the amount of material goods per additional unit of emission abatement is reduced. Compared with the starting situation, each emission reduction target would then be compatible with a higher level of material goods production.

[41] Goulder and Schneider (1999, p. 218).

[42] See Sect. 5.1.

[43] See Sect. 5.1.

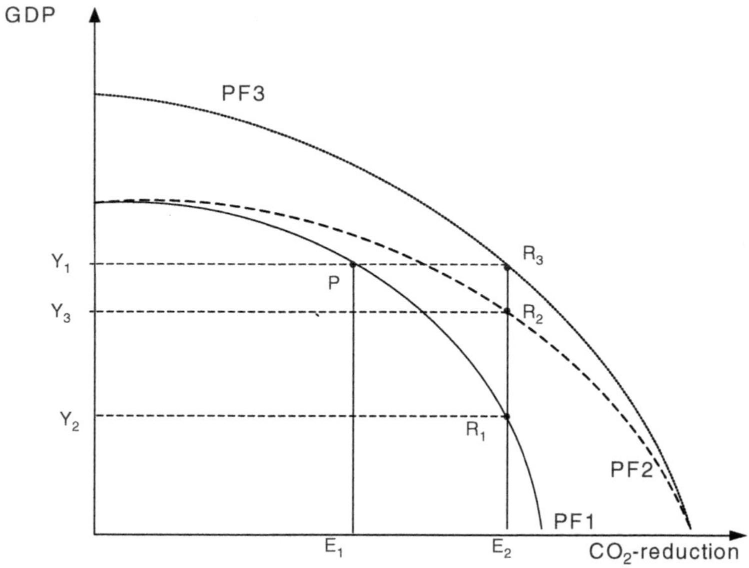

Fig. 2.3 Effects of technical change on the production frontier

climate protection because, according to the results of energy system analyses, high reduction capacities and clear, additional financial burdens are expected, especially in this sector. For the assessment of the direct additional costs, it is essential to know, in spite of the considerable significance of standard-based instruments in this sector, whether a *policy-induced technological change* can be reckoned with.[44]

In the theoretical analysis, the effects of a policy-induced technical change can be explained using the well-known concept of the production frontier (Fig. 2.3). In order to reach a predefined CO_2 reduction goal E_2, it is necessary to move on the production frontier from P to R_1. This leads to a reduction of GDP from Y_1 to Y_2. In Fig. 2.3, it is assumed that technical change is induced by a climate policy. This leads to a shift of the production frontier from PF1 to PF2. Thus, point R_2 on the new production frontier allows CO_2 emissions to be reduced by the same amount, but with less loss in GDP.

2.3.2 *Productivity Effect of Investments in Climate Protection*

Technical change in many cases is linked to investments having been made. New systems incorporate technical change and bring about a modernisation of the capital stock. The production possibilities of a national economy increase over time due to

[44] See Sect. 5.2 which deals with empirical results of climate policy induced technical change.

the growth and renewal of the capital stock. It has to be asked which impacts the *diffusion of climate protection technologies* – driven by a climate policy – have on this process. Here it is decisive whether climate protection technologies themselves show a productive effect in the sense of increasing the material goods output potential. This argument is also summarised in Fig. 2.3. If the investment in CO_2 reducing technologies has no productivity effects, this results in a shift of the production frontier from PF1 to PF2.[45] However, if such an investment also has productive effects at the same time, the production frontier shifts towards a curve such as PF3 instead. It is obvious that in such a case the economy is better off, because PF3 compared to PF2 allows any reduction in CO_2 emissions to be reached with a higher level of GDP.

However, the argument is more complex than this because it also has to be accounted for that climate protection investments crowd out other productive investments. Under the assumption of a constant total investment volume, the following two cases are conceivable:

- In the first case, it is assumed that climate protection does not show any productive impact. However, in case of a climate policy, total investments consist of both productive and climate protection investments. By investing in productive technologies, the production frontier is shifted towards a higher production of goods, but does not bring about any reduction in CO_2 emissions. According to this case, the investments in climate protection technologies do not have any productive impacts, i.e. they only reduce CO_2 emissions. Together, both types of investment result in a shift of the transformation curve towards higher goods production and lower CO_2 emissions. However, since the climate protection investments do not have any productive impacts themselves, under the ceteris paribus assumption of a constant investment volume, the productive investments of companies are then crowded out. Thus, the increase in productivity is lower compared to the development in which all investments are used for productive technologies. To sum up, in case 1, the macroeconomic productivity increase would be diminished by such a "technological crowding out".
- In the second case, again, both traditional productive investments and investments in climate protection are made. The same effect occurs for the traditional investments as in the first argument. The differences are to be found in the character of the climate protection technologies. Here, it is assumed that they also have a productive character. They thus increase – in contrast to the first argument – simultaneously the production possibilities of material goods. This effect occurs, for example, if the climate protection technologies represent new efficient production technologies which replace older production technologies burdened with higher emissions and lower productivity.[46] The crowding out of

[45] Such an assumption was implicitly made in the arguments about induced technological change in Sect. 2.3.1.

[46] Xepapadeas and de Zeeuw (1999, p. 167) refer to this as the "modernisation effect".

investments with productive effects derived under the ceteris paribus condition of a constant investment volume is then alleviated, or, in an extreme case, does not occur at all.[47]

The assumption of a constant investment volume can be abandoned if it is assumed that there is an *increase in the investment volume*.[48] Under this assumption, if climate protection investments have a productive character, this would be tantamount to a "technological crowding in" and an increased modernisation of the national economy would follow in its wake. This kind of effect takes place, e.g. if older production systems lose profitability due to the introduction of an eco-tax, and are taken out of service earlier than planned. This induces additional investments in new, more productive systems with lower emissions.[49]

The effects of technical change induced by the climate policy depend very strongly on which of the two hypotheses with regard to the productive impact of climate protection technologies is given greater weight. The hypothesis of a non- productive effect of investments in environmental protection is probably valid for end-of-pipe solutions which are added on to the production systems and tended to dominate environmental protection in the 1970s and 80s. On the other hand, it seems plausible that those investments which directly affect production (production-integrated environmental protection), and which have become more important, have more productivity-increasing effects than the end-of-pipe systems. First empirical results indicate that climate protection investments do indeed have a productive effect as well.[50] However, it is also clear that the magnitude of this effect depends on the technology, and that, in general, a substantial increase in productivity is induced only by some of the climate protection investments.

2.3.3 *First Mover Advantage*

Besides price competitiveness, which is influenced by cost effects, foreign trade successes are also determined by *quality competitiveness*. Above all for technology-intensive goods, which include climate protection technologies, high market shares depend on the innovation ability of a national economy and its early market presence. If there is a forced national strategy to reduce greenhouse gas emissions, these countries tend to specialise early in the supply of the necessary goods. If there is a

[47] This is the effect if climate protection investments result in the same increase in productivity as new productive investments.

[48] This kind of increase in investments could either take place at the expense of consumption or, in a Keynesian argumentation, be additionally induced (see Sect. 2.3.4).

[49] Xepapadeas and de Zeeuw (1999, pp. 173–174).

[50] Walz (1999). However, if a CO_2 reduction strategy moves towards separation and storage of CO_2, such an effect would not occur.

subsequent expansion of the international demand for these goods, these countries are then in a position to dominate international competition due to their early specialisation in this field.[51]

Being able to realise these kinds of first mover advantages requires other countries to follow suit. Given the growing demand for energy on the one hand, and the pressure to push for non-fossil fuels on the other, there is a high probability of this taking place. For first mover advantages to be realised, however, the domestic suppliers of climate protection goods have to be competitive internationally so that they themselves and not foreign suppliers meet the demand induced by the domestic pioneering role.[52] Taking the globalisation of markets into account, this requires that competence clusters are established which are difficult to transfer to other countries with lower production costs. These competence clusters must consist of high technological capabilities linked to a demand which is open to new innovations and horizontally and vertically integrated production structures. The following factors have to be taken into account when assessing the potential of countries to become a lead market in a specific technology:

- Lead market capability: It is not possible to establish a lead market position for every good or technology. One prerequisite is that competition is driven not by cost differentials alone, but also by quality aspects. This is especially valid for knowledge-intensive goods. In general, the technology intensity of climate protection technologies can be judged as being above average or even (e.g. photovoltaics) high tech. Other important factors are intensive user-producer relationships and a high level of implicit knowledge. These factors are not easily accessible to competitors, difficult to transfer to other countries and benefit from local clustering.[53] Two other important characteristics are high innovation dynamics and high potential learning effects. They are the key that a country which forges ahead technologically is also able realizing a degression in costs.
- Competitiveness of complementary industry clusters: Learning effects are more easily realised if the flow of (tacit) knowledge is facilitated by proximity and a common knowledge of language and institutions. The results of Fagerberg (1995b) can be explained in this way. He found strong empirical evidence that the international competitiveness of sectors and technologies is greatly influenced by the competitiveness of interlinked sectors. By and large, climate policy related technologies have very close links to electronics and machinery. Thus, it can be argued that countries with strong production clusters in these two fields have a particularly good starting point for developing a first mover advantage.
- The importance of the demand side can be traced to the work of economists from the 1960s.[54] There are various market factors which influence the chances of a

[51] Porter and van der Linde (1995, pp. 104–105), Taistra (2001, pp. 242–243). See also Blümle (1994).

[52] Ekins and Speck (1998, pp. 42–43), Taistra (2001, pp. 250–251).

[53] See Kline and Rosenberg (1986), Lundvall and Johnson (1994), Asheim and Gertler (2005).

[54] E.g. Linder (1961); see Hippel (1986), Porter (1990), Dosi et al. (1990).

country developing a lead market position.[55] In general, a demand which is oriented towards innovations and readily supports new technological solutions benefits a country in developing a lead market position. Another factor is a market structure which facilitates competition. The price advantage of countries is very important. Countries increasing their demand fastest are most able to realise economies of scale and learning effects. If one looks at the diffusion rate of the various forms of climate protection technologies in different countries, it can be seen that European countries have been forging ahead recently. Furthermore, the political goals for the EU will bolster this advantage in future. Nevertheless, there are also other countries which have recently increased their diffusion rates. If large markets, such as the U.S., China, India or Brazil, increase their use of mitigation technologies, this will cause a huge rise in absolute numbers which might also strengthen their price advantage.

- In addition to technological and market conditions, a lead market situation must also be supported by innovation-friendly regulation.[56] This is especially true for sustainability innovations in infrastructure fields such as energy, water or transportation. In these fields, the innovation friendliness of the general regulatory regime, e.g. with regard to IPR or the supply of venture capital, must be accompanied by innovation-friendly sectoral and environmental regulation resulting in a triple regulatory challenge.[57] There is a lot of additional research necessary to develop a clear methodology on how to analyse the innovation friendliness of regulation. One promising approach is a heterodox one which uses the sectoral systems of innovation approach as guiding heuristics. The first empirical case studies for renewable energies show how such an approach can be operationalised.[58] This approach also offers the opportunity to combines various paradigms. The effect of different instruments on innovation is a key question analysed within the neoclassical environmental economics paradigm. Other paradigms contribute to this question, e.g. transaction and evolutionary economics, which emphasise take a somewhat different look at decision making. They state that the decisions, e.g. with regard to financing renewable energy technologies, follow a different paradigm (e.g. other valuation of financial risks, bounded rationality with regard to alternative suppliers of electricity). Furthermore, the policy analysis approach of political scientists emphasises the long-term character of political goals, or the comparatively important role of green policies for voters, which are key supportive context factors favouring innovations.
- Since the Leontief Paradox and subsequent theories such as the Technology Gap Theory or the Product Cycle Theory, it has become increasingly accepted that international trade performance depends on technological capabilities.[59] This has been supported by recent empirical research, which underlines the

[55] Beise (2004), Beise and Cleff (2004).

[56] Blind et al. (2004).

[57] See Chap. 8.

[58] See Chap. 8.

[59] Posner (1961), Vernon (1966), Fagerberg (1994), Wakelin (1997), Archibugi and Michie (1998).

importance of technological capabilities for trade patterns and success.[60] Thus, the ability of a country to develop a first mover advantage also depends on its comparative technological capability. If one country has performed better in the past with regard to international trade than others, it has obtained key advantages on which it can build future success. Thus, trade indicators such as shares of world trade or specialisation indicators such as the Relative Export Advantage (RXA) or the Revealed Comparative Advantage (RCA) are widely used to compare the technological capability of countries. Furthermore, a country has an additional advantage in developing future technologies if it has a comparatively high knowledge base. Thus, patent indicators such as share of patents or the Relative Patent Advantage are among the most widely used indicators to measure technological advantages.

Empirical findings to support this hypothesis can be drawn from studies of trade relations using indicators based on patent or trade data.[61] For both types of indicators, the share of the most important countries at the world total was calculated (patent share, world export share). Furthermore, specialisation indicators (relative patent advantage (RPA); relative export advantage (RXA) and revealed comparative advantage (RCA) were calculated, in order to analyse whether or not the countries specialise on the climate protection technologies. They were formed in a way that the indicator shows values between −100 (extremely weak specialisation) and +100 (extremely strong specialisation):

- Relative patent advantage: for every country *i* and every technology field *j* the RPA is calculated according to

$$\text{RPA}\,ij = 100 * \tanh \ln\,[(\text{p}\,ij / \sum_i p_{ij}) / (\sum_j p_{ij} / \sum_{ij} p_{ij})]$$

- Relative export advantage: for every country *i* and every technology field *j* the RXA is calculated according to

$$\text{RXA}\,ij = 100 * \tanh \ln\,[(\text{ex}\,ij / \sum_i ex_{ij}) / (\sum_j ex_{ij} / \sum_{ij} ex_{ij})]$$

- The revealed comparative advantage includes both exports and imports into the analysis and is calculated for every country *i* and every technology field *j* according to

$$\text{RCA}\,ij = 100 * \tanh \ln\,[(\text{exij}/\,\text{imp}\,ij) / (\sum_j ex_{ij} / \sum_j imp_{ij})]$$

[60] Wakelin (1997), Fagerberg (1995a), Fagerberg and Godinho (2005), Blind and Frietsch (2005).

[61] For the relevant indicators, see Legler et al. (1992, pp. 89–93) and Grupp (1998), who assign the RCA in particular high significance for measuring technologically-determined foreign trade advantages.

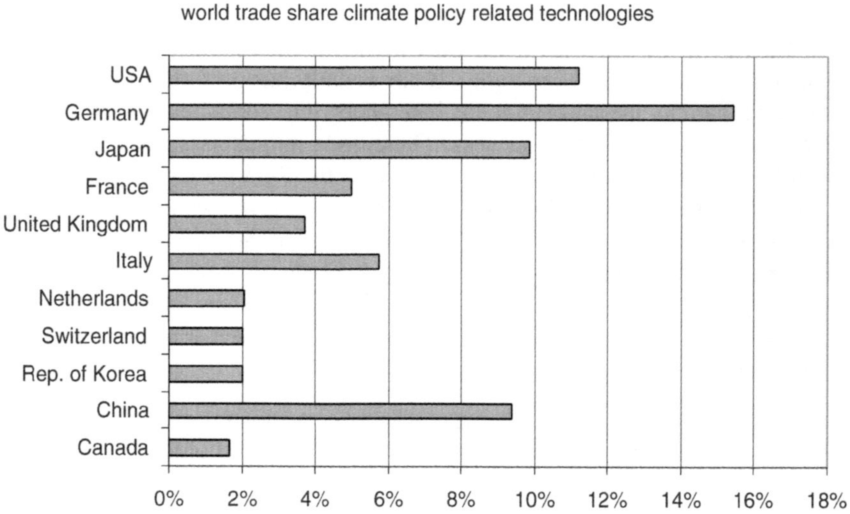

Fig. 2.4 World export shares of climate policy-related technologies in 2005 (Data: calculation of Fraunhofer ISI, Karlsruhe)

Climate protection technologies are neither a patent class nor a classification in the HS-2002 classification of the trade data from the UN-COMTRAD databank which can be easily detected. Thus, for each technology, it was necessary to identify the key technological concepts and segments.[62] They were transformed into specific search concepts for the patent data and the trade data. This required an enormous amount of work and substantial engineering skills. Furthermore, there is a dual use problem of the identified segments, and some segments – especially in the trade data – do not necessarily indicate that the technology is sustainable. In order to reflect that ambiguity the term climate relevant technology is used.

The importance of exports of climate policy related technologies can be seen from Fig. 2.4. World exports of climate policy related technologies are dominated by Germany, the U.S. und Japan. Furthermore, the other big EU countries play an important role. However, there are also new exporting countries entering the game, notably China and South Korea. Thus, it is very important to look at the technological basis behind these exports in the various countries.

The patent analysis reveals that climate policy related technologies have a considerable innovation dynamics. Between 1991 and 2004, the annual patent application in this field increased by 250%. The most important countries are the U.S., Germany, and Japan. However, over time, the share of the U.S. is shrinking. Germany's share remains largely unchanged, whereas Japan's share has been is increasing steadily (Fig. 2.5).

[62] This work extends the analysis of Legler et al. (2006) and DIW/ISI/Berger (2007) further by including additional climate policy related technologies into the analysis, and by moving from a stronger EU/OECD country oriented methodology towards including the total world market.

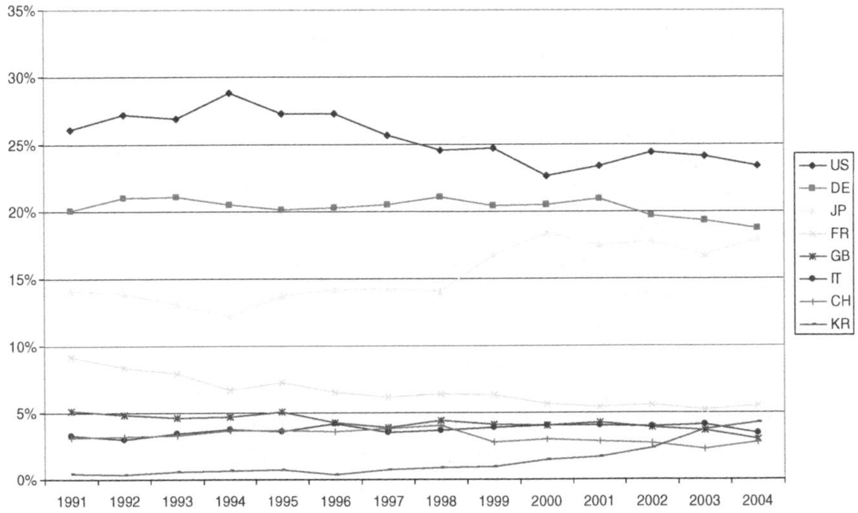

Fig. 2.5 Development of world patent shares of climate policy-related technologies (Data: calculation of Fraunhofer ISI, Karlsruhe)

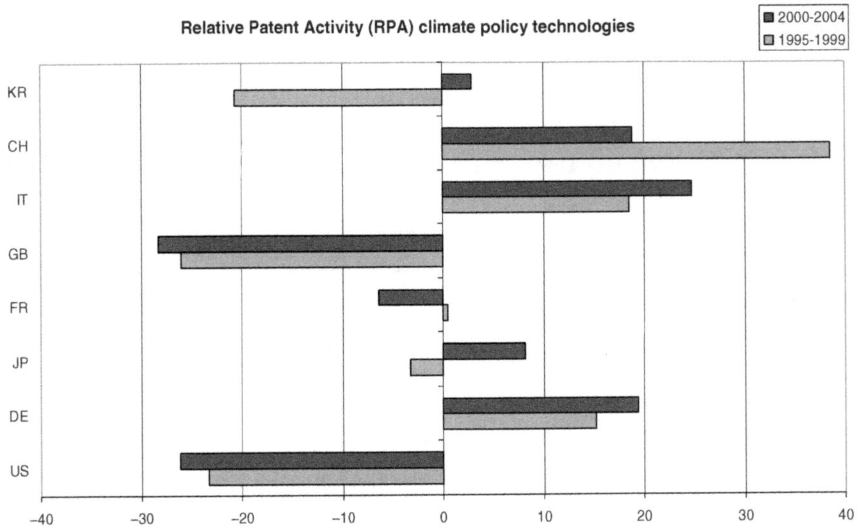

Fig. 2.6 Relative patent activity of climate policy-related technologies (Data: calculation of Fraunhofer ISI, Karlsruhe)

The shares at the absolute numbers do not account for the fact that the countries differ in size. Thus, in addition, specialisation measures are used which indicate whether or not a country is specializing on the technologies (Fig. 2.6). The numbers clearly indicate that Italy, Switzerland and Germany are very strong in the climate related patenting, However, in the last years, Japan has been able to specialise in this technology field too.

Another specialisation measure is related to the trade data itself. The most comprehensive indicator is the revealed comparative advantage. In addition to exports, it also takes the imports into account. A positive value indicates that the country has been specializing on the analysed goods, and vice versa. Figure 2.7 gives the results of the RCA for climate policy related technologies.

The data show that climate policy related technologies clearly form a very successful segment of the traditional industrialised countries. Germany, Japan, and even the US are showing positive RCA values. However, the data is not without caveats, especially if one looks at disaggregated data for single technologies. The RCA is difficult to interpret if imports are influenced by rapidly growing demand and domestic constraints to keep up capacity growth with increasing demand. The resulting surge in imports – and orientation of domestic producers on the home market – drive the RCA to negative level, even though the country might be very competitive. Such a situation has been taking place for wind energy in Germany, where demand outstripped national production capacities in the early 2000s. The latest data indicates, however, that the situation has been changing with regard to wind energy. Germany now not only holds about 30% of the patent applications in this technology field, but has been reaching positive RCA values too. Nevertheless, with regard to solar cells for photovoltaic, Germany is still having a negative RCA, due to an import surge from countries such as Japan. In other technological fields,

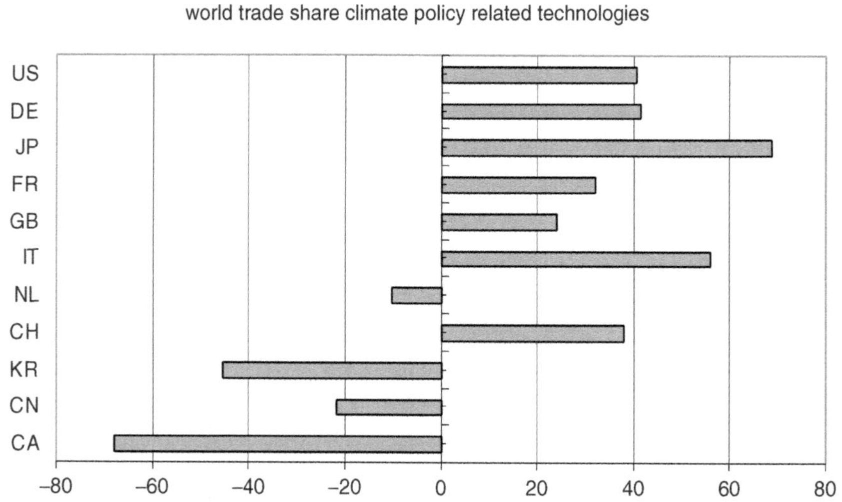

Fig. 2.7 Revealed comparative advantage of climate policy-related technologies in 2005 (Data: calculation of Fraunhofer ISI, Karlsruhe)

Germany has obtained positive RCA values for substantial times. This holds especially in the field of energy efficiency technologies, but also in supply oriented technologies such as carbon capture and storage (CCS).

Overall, it can therefore be assumed that Germany could profit from a first mover advantage in climate protection goods. However, it has not yet been possible to quantify more accurately the extent of such an effect, despite of all the progress with regard to measuring technological capability of the technologies. First, the measurement of the demand factors seems to be very case specific and stress the significance of demand conditions which are difficult to generalise. Secondly, and even more important, is the very high importance of an innovation friendly regulation especially for climate policy related technologies. Measurement of the intensity of the policy intervention – necessary for statistical analyses – is extremely difficult. Indeed, there is already significant disagreement on the mechanisms and the direction of influence of regulation on innovation, which underline the importance of additional research in that area.[63]

2.4 Demand Effects

There are two types of demand effects to be considered: First, the effect of structural changes between the components of final demand, and second the effect of an increase in aggregate demand on macroeconomic variables such as consumption and investment.

2.4.1 Structural Changes

A structural change in the component of final demand occurs if the climate policy triggers an increase in investment. Implementing a climate policy requires additional investments to increase mitigation capacities (direct positive impulses). At the same time, there is a drop in demand for both conventional energy carriers and conventional energy supply investments (direct negative impulses). With the exception of the case of a no-regret-potential, the costs for a climate policy are assumed to be higher than the capital and running costs for the conventional energy supply. Typically, a substantial share of the higher costs is transferred to the consumers. Thus, they have less income to spend on other consumer goods. Another possibility is the loss of tax revenues caused, for example, by tax exemptions for climate friendly technologies. Government then has to reduce other expenditures or increase revenues and will thus crowd out other investments or consumer spending. To sum up, compensatory effects occur within the structural adjustment mechanism

[63] See Chap. 8.

and, in the case of higher costs, negative consumption effects have to be accounted for when analysing structural shifts of demand.

Since numerous inputs from other sectors are necessary to supply the respective demand, the direct positive and negative impulses are carried forward as positive and negative *indirect effects* according to the production linkages of the industries involved. Thus the different positive and negative impulses lead to a different structural composition of the overall economy.

The argument so far has demonstrated the importance of the positive and negative demand impulses. It has been shown that it is the effect on the supply chain which influences the structural demand effects. They include the effects due to interlinkages between the production sectors. A more formal analysis reveals the following with regard to the overall effect on employment: The total production induced by an impulse is the sum of sectoral production p_k in all sectors *k*. The total employment which is induced by an impulse depends on the total domestic production in each sector in the value chain, and the labour intensity in each of these sectors. Furthermore, the total domestic production in each sector equals the overall total production of goods of each sector *k* minus the imports of each sector *k*. Thus, the total employment effect of an impulse can be written as:

$$employment = \sum_{k=1}^{K} p_k * (1 - import\ ratio\ k) * (labour\ intensity\ k) \quad \forall k = 1, \ldots, K$$

- The average import ratio, calculated for the complete value chain of an impulse, demonstrates which percentage of total production induced by the direct impulse is imported. The higher the import ratio, the lower the domestic production.
- The average labour intensity, also calculated for the complete value chain, demonstrates how many persons are employed per Euro of total domestic production induced by the direct impulse.

Thus, by comparing the labour intensities and the import intensities of the value chains of the positive and negative impulses of a climate policy, it is possible to get a first impression of the structural effects on employment. For energy-importing countries, it is significant that a considerable share of the negative demand effects – namely the reduction in demand for imported energy – takes effect not domestically but in the energy-producing countries. If a higher share of the climate policy investments is produced domestically, a net increase in domestic production results. If, in contrast, a considerable share of the energy is produced domestically, and a considerable share of the investments to reduce traditional energy consumption has to be imported, then a reduction in aggregated domestic demand results.

Figure 2.8 gives a first impression about the order of magnitude of import ratios of the value chain for impulses from different sectors and EU countries:[64] The average import intensity of the mineral oil product chain is by far the highest. On the

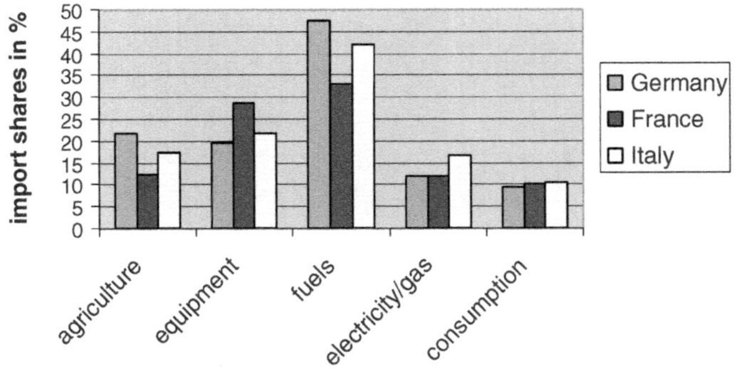

Fig. 2.8 Import shares of the complete value chain of various goods (Data: calculation of Fraunhofer ISI, Karlsruhe)

other hand, the total accumulated import of the value chain of electricity production is rather low. This reflects, among others, the important role of very capital-intensive nuclear power in France and Germany and the importance of German-based lignite in electricity production. The import shares of the value chain of average consumption are also quite low. They are an important indicator for the effect of impulses arising from the need to compensate for the additional costs. The value chains of the sectors most likely to benefit from climate policy strategies, e.g. investments in equipment or the agricultural and forestry sector (use of biomass) tend to have import shares which are in-between the value chains of the sectors they will substitute. Given these results, a substitution of conventional electricity production and oil products by renewable energy and more rational use of energy has no clear effect with regard to import substitution and depends on the specifics of the climate policy strategy which influences the composition of technologies.

The effect of structural changes in demand on labour intensity has to be accounted for, especially with regard to employment. An increase in employment results if the value chains of the sectors favoured by the climate policy have higher labour intensities than the value chains of the sectors favoured by the conventional energy supply. Typically, high labour intensities can be observed in the agricultural and forestry sectors. These result in an above average labour intensity of the associated value chains (see Fig. 2.9). The value chain of fuels production has low labour intensity, followed by the value chain of conventional electricity production. The

[64] These results are based on calculations performed with the international ISIS model which is based on the I/O-tables for various European countries. The author thanks his colleague Philipp Seydel from ISI for performing the model runs.

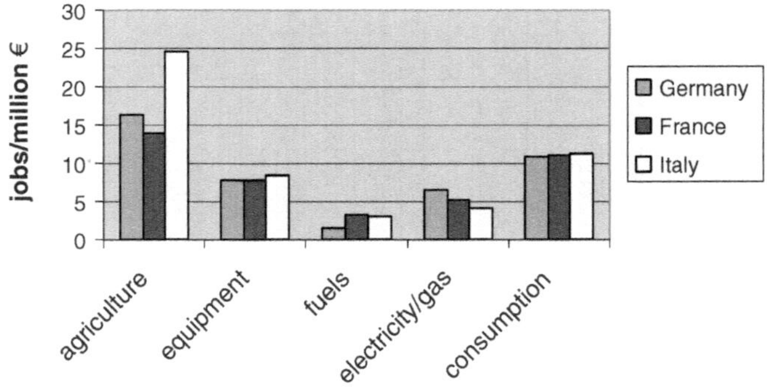

Fig. 2.9 Labour intensity of the complete value chain of various goods (Data: calculation of Fraunhofer ISI, Karlsruhe)

labour intensity of the investment sectors, by and large, is in between the labour intensity of fuels and electricity production. Thus, it can be assumed that the substitution of conventional energy supply by climate policy strategies generally leads to a modest increase in labour intensity. However, if the additional cost of the climate policy is very high, the value chain of consumption goods becomes increasingly important, because more and more of consumption must be sacrificed to cover the additional costs. The labour intensity of the value chain of consumption is above the one of the value chain for equipment. Thus, the effect of structural change towards labour intensive sectors becomes less prominent, the higher the cost difference of renewable energies to conventional energy supply.

2.4.2 Income Multiplier and Accelerator Effects

Demand-side effects are the cornerstone of the Keynesian model, which sees unemployment as caused by a deficit in aggregate demand. Assuming that the conditions for Keynesian unemployment are met, positive growth and employment effects are to be expected if climate policies result in an increase in the effective demand for goods.[65] This can lead to a self supporting increase in business activities triggered by mutually reinforcing income multiplier and accelerator effects. These effects depend on the economic conditions and the assumed reactions of the actors which are at the centre of the debate on Keynesian economics. An important assumption is that the demand from climate policies does not crowd out other segments of aggregate demand.

---

[65] For a theoretical presentation of environmental policy in the Keynesian model, see Lintz (1992, pp. 42–47).

However, some limitations have to be taken into account when considering this argument. The effect of Keynesian demand policy in the overlapping area of rational expectations, international goods and financial markets is substantially more complex than the mechanistic description above may suggest. Thus, the chances of success of a demand policy have been regarded with scepticism for some time.[66] It also has to be questioned whether the volume of demand changes to be moved by the climate protection would be sufficient for more than marginal effects of the business cycle. Limitations also result from the temporal links of the climate policy with a demand policy. A climate policy has to be designed for the medium to long term. In contrast, a demand policy has to be oriented on macroeconomic constellations. Only if these are favourable to a demand policy, can a climate policy induce positive macroeconomic effects in line with the impact mechanism described here.[67] On top of this, it is unclear whether the effects achieved due to an increase in demand can definitely be assigned to the climate policy. There would be no reason to mobilise the demand-increasing effects of the climate protection policy for employment policy reasons if the assumed potentials of a demand policy had already been tapped by other measures. It can be stated that climate policy cannot and should not be a substitute for other instruments of business cycle policy. Independently of this, it may give expansionary impulses which, depending on the macroeconomic constellation, bring about favourable effects, particularly since they might take effect virtually unnoticed by the formation of expectations under the disguise of environmental protection.

2.5 Combined Effects of the Impulses

In the previous sections impulses were discussed which are triggered by the various effect mechanisms. If one looks at the direction in which they work, a somewhat contradictory picture emerges (see Fig. 2.10):

- The cost and price effects resulting from the realisation of costly CO_2 reduction potentials have a clearly negative effect on the macroeconomic targets. These impulses are represented by a curve of type K_1 in Fig. 2.10.
- This picture becomes more sophisticated if, in addition, the existence of inefficiencies is considered. If a no-regret potential or a strong double dividend is assumed to exist, impulses result as shown in curve K_2. Caused by the increases in efficiency, a cost decrease results at first for the reduction of greenhouse gas emissions which has a positive impact on the macroeconomic targets. A neutrality of the cost impulse occurs if the total cost savings of the economic measures are equal to the total cost increases of the uneconomic ones. In the graph, this

[66] Landmann (1984, pp. 211–212).

[67] Conversely, it would be counterproductive for climate protection if climate policy had to be damped down in order to cool down an economy.

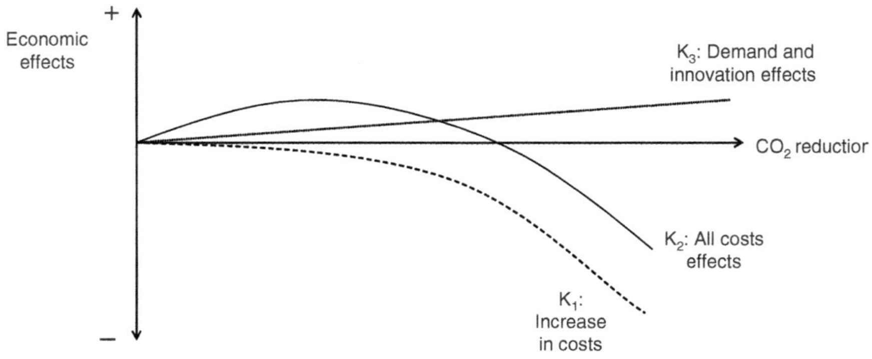

Fig. 2.10 Isolated effects of the impulses

corresponds to the point of intersection S of curve 2 with the abscissa. Up to this reduction of greenhouse gases, positive macroeconomic impulses emanate from the cost effects, only CO_2-reductions beyond this point trigger negative impulses in analogy to the above argument.

- The effects on the aggregate demand triggered by climate protection measures must be additionally taken into account. If one assumes a situation of underemployment corresponding to Keynesian perceptions, a demand impulse induced by climate protection measures is reinforced by the income effects. Under these assumptions, the effect of a climate protection policy can be symbolised by a straight line, K_3. Here, it is disregarded that the demand impulse could take place in a phase of the business cycle which is not appropriate for the given macroeconomic situation and, e.g. would trigger a wage-price-spiral or reactions of the central bank.
- In Sect. 2.3, it was argued that a climate policy could result in increasing productivity, increased generation of new technological solutions and the realisation of first mover advantages. The straight line K_3 in Fig. 2.10 symbolises the effect on national economy under the assumption that the positive innovation effects dominate the negative ones. However, it should be added that more detailed empirical analysis is required of both the existence and/or strength of these impulses.

It can be ascertained that the total impact results from the *interaction* of the various *mechanisms* and cannot be derived from the isolated observation of individual sub-effects. The intersection of curve 2 with the abscissa makes clear that the direction can change i.e. that moderate climate protection may bring about positive impacts, whereas there may be negative impacts from a very forced/accelerated climate protection. Another main point is that the direction of the effect not only depends on the climate protection policy itself, but also partly on how it is *embedded in the economic policy*. This is obviously the case for an energy tax, the effect of which on the national economy depends very heavily on the utilisation of the tax revenue and the interaction with the existing tax system. The size and direction of the income cycle effect also depends to a considerable extent on aspects external to the climate policy such as, e.g. whether a Keynesian under-employment situation exists, or the reactions of the bargaining parties and the central bank. Depending on the economic policy framework conditions, one and the same climate policy may thus have a different macroeconomic effect. Conversely, it can be argued that it is not actually the climate policy which generates these effects, but the economic policy per se. The climate policy here only acts as an additional motivation for implementing these measures – albeit as a motivation which is justified in itself, namely climate protection.

The arguments put forward so far concerned the isolated effects of the different economic mechanisms. Taken together, the combination of the different effects leads to a situation characterised in Fig. 2.11: Up to a certain point, a climate protection policy is likely to result in an increase in production. However, if more than a modest reduction of CO_2 emissions is aimed at, the negative effects become stronger and stronger leading to losses in production. The effects on employment are similar. However, if tax policies are used, with revenue being applied to lower the cost for labour, or if structural demand effects work in favour of more labour intensive sectors, the positive effects on labour demand are stronger and the negative effects start to prevail at a higher reduction level.

For policy making, the key aspect is how relevant the level of effects is from a political perspective. The following questions are particularly interesting:

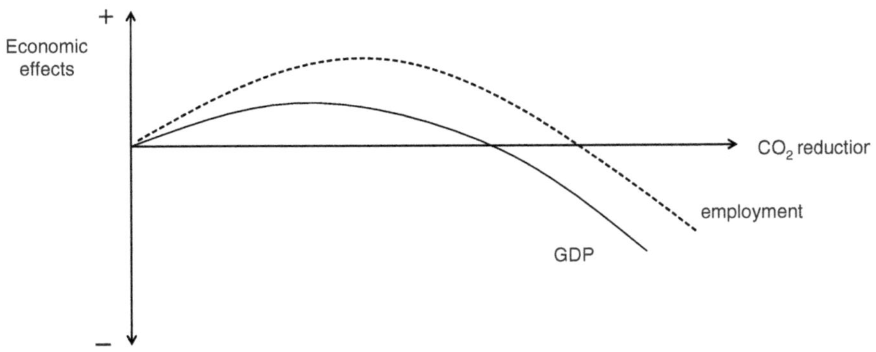

Fig. 2.11 Combined effects of the impulses

- At which level of CO2 reduction do positive effects occur; how big are they?
- Up to which level of CO_2 reduction are no severe macroeconomic losses to be expected?
- How big is the increase in employment if, for example, a CO_2 reduction is achieved which has no effect on production?

The theoretical analysis does not make any predictions about these kinds of questions. Thus, it is necessary to turn to the empirical macroeconomics of climate change.

Chapter 3
Empirical Results of the Macroeconomic Impacts

This section looks at empirical macroeconomic studies of the economic impacts of climate protection policy. Based on a characterisation of the available *modelling approaches*, an overview is given for Germany. The results obtained using the various modelling approaches are compared with each other in order to derive an overall assessment of the macroeconomic results.

3.1 Modelling Approaches

To empirically assess the effects triggered by climate policy, it is necessary to use empirical models which quantify the various links between the different economic mechanisms. Those used include primarily *macroeconometric and computable general equilibrium models (CGE)* but also some *input/output models*. There are specific strengths and weaknesses associated with these different types of model.[1]

In *CGE models*, price changes trigger adjustment reactions in all other sectors. *Supply and demand functions* are used for clearance of the markets which are balanced by the price mechanism. The microeconomic foundation and the long-term orientation of the model achieved by mapping economic feedback processes are seen as the main advantages of this model. On the other hand, the empirical foundation of *equilibrium models* is criticised, since the *calibration* of many structures and parameters is not based on time series, but is usually conducted for one particular year. On top of this, the functional forms are generally limited to constant elasticity of supply *(CES) production and utility functions*. At the same time, many parameters have to be entered exogenously. Due to the assumption of efficient markets, on which the models, in general, are based, there are also major difficulties with the

[1] A review of the pros and cons of the individual models for analysing environmental policy is given by West (1995), Pfaffenberger (1995), Fankhauser and McCoy (1995), Destais (1996), Bhattacharyya (1996), Barker and Johnstone (1998), Conrad (1999), and Duchin and Steenge (1999).

R. Walz and J. Schleich. *The Economics of Climate Change Policies*.
Sustainability and Innovation,

portrayal of no-regret potentials. Thus, together with the emphasis placed on cost effects, the model structure makes it difficult to take increases in efficiency into account. These leads to a tendency that climate policies result in negative effects if they are analysed using these models: "To some extent, however, the negative results arise by definition. In effect the competitive losses estimated by CGE models are merely incidental manifestations of more fundamental characteristics of the models themselves."[2] This criticism is alleviated by the fact that the existence of distortionary taxes is integrated in the model and therefore, in principle, double dividend effects can also be illustrated.[3] Nevertheless, the overall conclusion is that the *CGE models* concentrate on the category of price and cost effects and, in doing so, exclude the existence of no-regret potentials.

Macroeconometric models illustrate the economic relationships based on both, definition and behavioural equations. Corresponding to the theoretical orientation of the models, *neo-classical or demand* aspects can be emphasised. The advantage of these models lies in the fact that very different macroeconomic effects can be illustrated as long as the necessary data to do so are available. At least in the short- and perhaps medium-term, they are assigned higher reality content than the *CGE models*. This has to be set against a comparably greater effort for model specification and adaptation. Furthermore, the functional forms used in some macoeconometric models are more simplistic than in CGE-models, which can lead to internal inconsistencies. The equations – which form the core of the models – are calculated using historical data. Especially for the application of long-term forecasts, it must be noted that the fundamental relations, which show up in the correlations calculated for the time period for which data is available, also have to be valid for the future, at least in their basic structure.[4] In principle, a *macroeconometric model* is also able to consider the inefficiencies reflected in the statistical connections. As long as the level of the inefficiencies can be measured with suitable variables, changes in inefficiency can also be taken into account. Typically this is the case for the various forms of taxation. Consequently, the models are also used when examining the existence of a double dividend. It is more difficult to model the realisation of a no-regret potential. A model specification would be necessary here to depict the extent of the inefficiency in the relevant variables. Overall, *macroeconometric models* tend to assess the effects of climate protection policy slightly less pessimistically than the equilibrium models.[5] This is particularly distinct in the Keynesian modelling approaches, which focus on demand effects since they assume an increase in aggregate demand caused by the jump in demand triggered by investments in climate protection.

---

[2] Barker and Johnstone (1998, p. 99).

[3] In addition, inefficiencies on the job market have recently been incorporated in some models. See Conrad (1999, p. 1073).

[4] Barker (1999, p. 416).

[5] A factor here is that the price elasticities of foreign demand as well as the factor mobility tend to be weaker in the econometric models than in the equilibrium models (Barker & Johnstone, 1998, pp. 100–105).

Input/output models are able to estimate the indirect production effects and the associated employment effects triggered by changes in final demand. However, they are not able to depict the demand repercussions brought about endogenously by the circular flow of income effects. Furthermore, they neglect substitution effects or the change in competitiveness due to price and cost variations, unless they are included exogenously into the demand impulses which drive the models. Analyses with input/output models are therefore valid primarily for those questions (e.g. the use of individual technologies) for which a high sectoral degree of disaggregation is necessary, and for which it can be assumed that the analysed measures will have no impacts on the aggregated flow of income. If the latter assumption does not apply, it will be necessary to calibrate the aggregate production results with other model results.

Alongside the selection of the economic model to be used, the foundation of the data input and the design of the reference scenario are also decisive criteria. From among the wide variety of possible combinations of model choice, data foundation and design of the reference scenario, two modelling approaches have become prevalent which have found their way into the literature under the terms "top-down" and "bottom-up analysis".[6]

The top-down analysis is characterised by the application of *macroeconomic models* without detailed energy scenarios. It models the impacts of a climate policy on the national economy and, at the same time, derives the change in emissions on this aggregated level. To do so, so-called *macroeconomic* – Economy–Environment–Energy – Models (E3 models) are used, which essentially consist of either equilibrium or econometric models. The impacts on emissions are modelled completely detached from any concrete technology; the demand for the production factor energy is derived from an (aggregated) production function or from statistical correlations. Substitution effects between the production factors have an effect here. Accordingly, these modelling approaches are primarily able to analyse changes in the relative prices. Depending on the changes in factor prices and the substitution elasticities between the production factors, the factor energy is substituted by the factor labour, if the former becomes more expensive and the latter cheaper. In the relevant analyses, a climate policy is typically modelled as an energy/CO_2tax, whose revenue is recycled in different variants.

The *elasticities of substitution* between the production factors are decisive for calculating emissions and costs with such top-down models. Ultimately they extrapolate the trends from the calibration of the model for 1 year or those observed in the past to the future.[7] Especially if leaps in technology can be expected,

[6]See Wilson and Swisher (1993), IPCC (1995, pp. 274–282), Hourcade and Robinson (1996), Krause (1996) and Faucheaux and Levarlet (1999, pp. 1132–1137).

[7]It must be kept in mind here that the reactions to changed energy prices observed in the past depend on numerous factors which have to be considered in a projection. For example, Jorgenson and Wilcoxen (1993, pp. 338–339) point to the fact that the effects of gradual but continuous price increases have to be judged quite differently to the oil price shocks occurring during the oil crises and the subsequent oil price cuts.

this results in a clear underestimation of the emission reductions or an overestimation of the energy tax rate necessary to achieve an exogenously given emission target and the costs associated with this. Correspondingly, these modelling approaches show deficiencies with regard to the coverage of new technologies.[8]

The assumptions made about the reference scenario form a second important factor which plays a particularly significant role in analyses examining the effects of achieving exogenous targets. Many *top-down analyses* assume an autonomous improvement of energy efficiency, i.e. the influence of technical progress is measured in the trend over time.[9] The size of the Autonomous Energy Efficiency Improvement (AEEI) factor is decisive for the amount of emissions in the reference scenario. In turn, the degree of emission reduction necessary to reach a political target over time depends on this and thus also the energy tax rate necessary to model the attainment of the target via substitution effects. The larger the necessary emissions reduction is, the stronger the economic effects. This implies that setting the level of autonomous technical progress in energy efficiency determines the extent of the economic effects.

A major criticism of top-down models is therefore also the consideration of technological change. The criticism refers both to the difficulties in setting a realistic "autonomous" increase of energy efficiency and the insufficient consideration of policy-induced technological progress. If policy-induced technological change is not taken into account in the model, costs of policy interventions will be overestimated, ceteris paribus. Another form of endogenous technological progress, which results from so-called learning-by-doing effects, implies investments in reduction measures at an early stage (Van der Zwaan et al., 2002; Goulder and Matthai, 2000). Even top-down models which allow for endogenous technical change such as Goulder and Schneider (1999) or Buonanno et al. (2003) and other models surveyed, for example, by Löschel (2002) or Carraro and Galeotti (2002), do not allow for a link to actual technologies responsible for the technological development. Similarly, Popp (2004, p. 743) criticises that "none of the existing models make use of empirical estimates on the nature of technological change to calibrate the model". Similarly, the study made by the German Advising Council to Green Accounting on the suitability of the German top-down models for illustrating the effects of environmental policy comes to the very restrictive result that "they did not manage to link the environment, economy and technology to the extent necessary to derive reliable results".[10]

In contrast to the top down approach, the second modelling approach (often referred to as bottom-up) starts from detailed energy scenarios which could also potentially include no-regret potentials. Typically, these scenarios are based on engineering-based partial models of the energy converting and using sectors, including different technologies and their improvement over time to capture all

[8] Frohn et al. (1998, p. 86).

[9] Meyer et al. (2001, p. 54). However, the efforts to improve the model have concentrated on that aspect recently.

[10] Frohn et al. (1998, p. 86); see also DIW/Fifo/RWI/ZEW (1996, p. 65).

energy saving possibilities. However, the situation is too complex to put forward a clear-cut distinction portraying bottom-up models as always including inefficiencies, and top-down models as neglecting them. Firstly, a reasonable number of bottom-up energy system models assume a purely optimising behaviour, i.e. they calculate the least-cost combination of a set of available or expected technologies for given production and emission targets. Hence, the modelled behaviour is in contrast to the arguments brought forward in Chap. 2 to support the existence of a no-regret potential. Secondly, including a decrease in inefficiencies in the models only enhances the validity of results if the assumptions on the magnitude of the efficiency increase are more valid than the assumption that there is no increase at all. Thus, there is a need for more empirical research on the magnitude of possible efficiency gains and on the factors triggering them.

In terms of portraying innovation effects, bottom-up model are able to relate technical change to actual technologies. In that sense though, technological change depends – to a large extent – on the set and the characteristics of the technologies included a priori in the database. However, some recent dynamic bottom-up models have started to allow for endogenous technological change via experience curves. The top-down modellers themselves concede that the bottom-up approach has considerable advantages with regard to considering technologies which have not yet been applied. "In any case, the quality of technology-based bottom-up studies and results must therefore be assessed as higher than those of macroeconomic top-down approaches."[11] However, *bottom-up* models tend to neglect market failures, transaction costs and uncertainty and, on their own, are not able to calculate directly the impacts on the national economy. The cost effects and also the direct positive and negative demand effects of the measures reviewed can, however, be derived from the scenario results. In the past, therefore, the results of bottom-up-models were used as data input for input-output analyses used to calculate the indirect demand effects induced by the production inter-linkages as well as the associated employment effects.[12] An evaluation of these kinds of studies states, for example, that for the ratios of the former Federal Republic of Germany at the end of the eighties/ beginning of the nineties, the net employment effects (i.e. taking the contractive effects in energy production and transformation into account) of rational energy use were around 100 additional jobs per petajoule of energy saved.[13]

[11] DIW/Fifo/RWI/ZEW (1996, p. 65) and Frohn et al. (1998, p. 87). However, the advantage of bottom-up models depends on the validity of the assumptions for future technology development incorporated into the models.

[12] See Garnreiter et al. (1983), Hohmeyer et al. (1985), Cames et al. (1996) and Laitner et al. (1998). Direct cost changes can be accounted for by translating them into compensating demand effects. Thus, if a climate policy leads to additional effects, this has to be modelled by reducing the demand for other goods in order to balance the cost increase.

[13] Jochem (1997a, p. 692). Due to rising productivities, this effect can be estimated at about 70 jobs per PJ at present. There are two reasons for the positive employment results: first, the higher labour intensity in the sectors benefiting from the policy (e.g. construction; see Chap. 4), and second, lower imports due to the substitution of imported energy by mostly German-produced capital goods. Thus, it becomes clear that this number cannot be easily transferred to other countries.

Due to the limitations of input/output models, this methodology cannot illustrate the repercussions on demand caused by the macroeconomic income effects or the numerous effect mechanisms which are triggered by cost and price changes, including rebound effects (Binswanger, 2001), i.e. that lower energy prices due to technological change will stimulate demand. As mentioned above, these kinds of analyses are able to provide meaningful statements primarily for mesoeconomic questions (e.g. the use of individual technologies). The core question of top-down approaches about the impacts of a revenue-neutral energy tax causes input/output models major problems since they are neither able to account for double dividend effects, nor the substitution effects between production factors triggered by the altered relative prices.

Overall, it is clear that all the models have considerable limitations. On the one hand, the spectrum of the economic effects examined by top-down models is much greater than in input/output analyses, but this is offset by the disadvantage of a more unrealistic modelling of the energy sector and the effects on emissions. At the same time, the economic models used in the top-down approaches only focus on part of the effect mechanisms, so that a concentration on specific effect mechanisms results, e.g. the effect of increases in costs or demand.

Against this background the question arises whether a reduction of the limitations could be achieved by linking the results of energy system analyses with macroeconomic models. If the potential advantages of such a combined approach are summarised, the following aspects result:

- Compared to the simple approach of using an input/output model, linking energy system models with a macroeconomic model makes it possible to take significantly more economic effect mechanisms into account. This can help to derive more realistic results.
- Compared to a top-down approach, a technology-specific empirical micro foundation of the main energy sector becomes possible through the combination with results from energy system analyses. In this way the much more detailed and resilient results of the effects of the climate protection policy on emissions can be considered and a more realistic level of policy intervention can be set with regard to the policy required to achieve exogenously given targets. This simultaneously lifts the restriction of the analysis to price instruments only. Combined bottom-up/top-down analyses can thus analyse complex action plans and do not have to restrict themselves to the hypothetical case of a climate policy consisting of only a CO_2 or energy tax.

Recent research efforts have started to incorporate technological aspects into the macroeconomic modelling of endogenous technological change (e.g. The Energy Journal, 2006). In most applications, selected technologies are incorporated at a rather aggregate level for the electricity sector in long-term endogenous growth models. At a more disaggregated level, Masui et al. (2006) link a global dynamic computable general equilibrium model and a bottom-up model for end-use energy technologies to analyse the effects of energy-saving investments on CO_2 emissions and the economy. Since the output of the bottom-up model is used as an input into the top-down model, the linking between technologies and macroeconomic variables is soft rather than integrated. In contrast, in Schleich et al. (2006) and Lutz et al. (2007) technological change in energy-intensive industry sectors is explicitly portrayed and linked to actual production processes within a macro-econometric model. Technology choice is then modelled via investments in new production process lines.

In spite of the obvious improvements which seem achievable by combining bottom-up and top-down approaches, certain restrictions cannot be overcome at present caused by the specification of the macroeconomic models or the limitations of the database. As for the difficulties in quantifying no-regret potentials, this is particularly true for the various innovation effects. Therefore it will be necessary in the near future to supplement model-based results with other additional studies and qualitative evaluations.

It can be stated in general that the empirical analysis of the economic impacts of climate protection is characterised by several competing models and modelling approaches. Parallel to this, but not completely congruent, the choice of model reflects the controversies already present when discussing the effect mechanisms. For this reason, the results of analysing the economic impacts will probably not be unambiguous. A sensible further development could be to link energy system analyses with macroeconomic models. Within the scope of policy analyses, however, developing an exact forecast is less important than identifying the scale of the effects and working out statements which are as robust as possible, i.e. are valid under different conditions. A critical comparison of the results compiled using the various approaches is therefore also necessary. This should present the main factors behind individual model results, reflect critically on each method with regard to its knowledge limits and, based on this, carry out an appraisal of the results with regard to generalisable conclusions including factors not able to be depicted in the model analyses.

3.2 Macroeconomic Effects

3.2.1 Review of the Results from Top-Down Approaches

Several national and some international surveys have examined the effects of a climate protection policy on production and employment in Germany. The majority of these studies concentrate on analysing the impacts of a CO_2/energy tax. The following studies are included in a comparison:

- The study drawn up by Welsch (1996) using the CGE model LEAN considers a moderate energy tax (somewhat more than 1 €/GJ). Its revenue is spent in different variants. Timeframe is the year 2020.
- The analysis on behalf of Greenpeace by DIW (1994) with the *econometric business cycle model* examines the effects emanating from a doubling of the energy prices (energy tax of Ð9 DM/GJ) within 10 years. Using the same model, DIW/Fifo (1999), in cooperation with the Public Finance Research Institute at

the University of Cologne (FiFo) on behalf of the Federal Environmental Agency, analysed the effects which ensue 10 years after the introduction of an environmental tax reform with an energy tax of 3 €/GJ.

- The study by RWI/Ifo (1996) on behalf of the BMWI examines a package of measures characterised primarily by short-term thermal insulation. The results described here refer to the year 2005.
- The analysis conducted by Meyer et al. (1997) using the sectorally disaggregated, econometric Panta Rhei model examines a comparatively intensive ecological tax reform with a tripling of the energy prices up to 2005.
- On behalf of the EU, Barker (1999) analysed the effects of a CO_2 tax of 18 €/t with the sectorally disaggregated econometric E3 model. The effects shown refer to Germany, the time horizon is 10 years.
- The study by Conrad and Schmidt (1999) commissioned by the EU and conducted parallel to Barker uses the GCE model GEM-E3 to analyse the effects triggered by a CO_2 tax of 22 €/t in the individual EU countries. The impacts given in the following statements refer to Germany. Using the same model, Schmidt and Koschel (1999) from ZEW analyse the impacts of a CO_2 tax of 105 €/t used to reduce the social security contributions. Both surveys have a time horizon of 10 years.
- Böhringer et al. (2001) examine the effects of a CO_2 tax of 35.2 €/t using a GCE model.
- The study set up by Prognos (2001) using an input-output model examines an energy policy action plan with a time horizon of 2020.
- In the context of the studies of DIW et al. (2001) and Frohn et al. (2003), the effects of the German eco-tax were analysed by Meyer et al. (2001, 2003), using the econometric Panta Rhei model, which was also used by Meyer et al. (1997). The results shown here refer to a continuation of the existing eco-tax to 2010, with additional increasing tax rates (3 cent/L for gasoline, and 0.5 cent/kWh for electricity) on top of the existing eco-tax and no special provisions for industry, compared with a reference case without any eco-tax.
- The study of Walz et al. (1995) analyses the effect of a policy mix in order to reduce the CO_2 emissions by 40% in 2020 compared to 1990. In contrast to the other studies, a combined bottom-up/top-down approach was used. Thus, for methodological reasons, this study is described in more detail in Sect. 3.3.2.

The main results of the single model analyses are summarised in Figs. 3.1 and 3.2. These list deviations in per cent compared to the absolute value of the reference scenario rather than differences in annual growth rates.

On the one hand, it is obvious that the majority of results indicate comparatively *low impacts*. It should therefore be emphasised that a reduction of the CO_2 emissions is much less significant macroeconomically than is surmised when looking at the exchanges in the political discussion. On the other hand, distinct differences do emerge between the individual studies. To some extent they can be explained by different modelling approaches, or by different policy measures being analysed which induce the effect mechanisms to a different degree. The following facts may help to interpret the individual results:

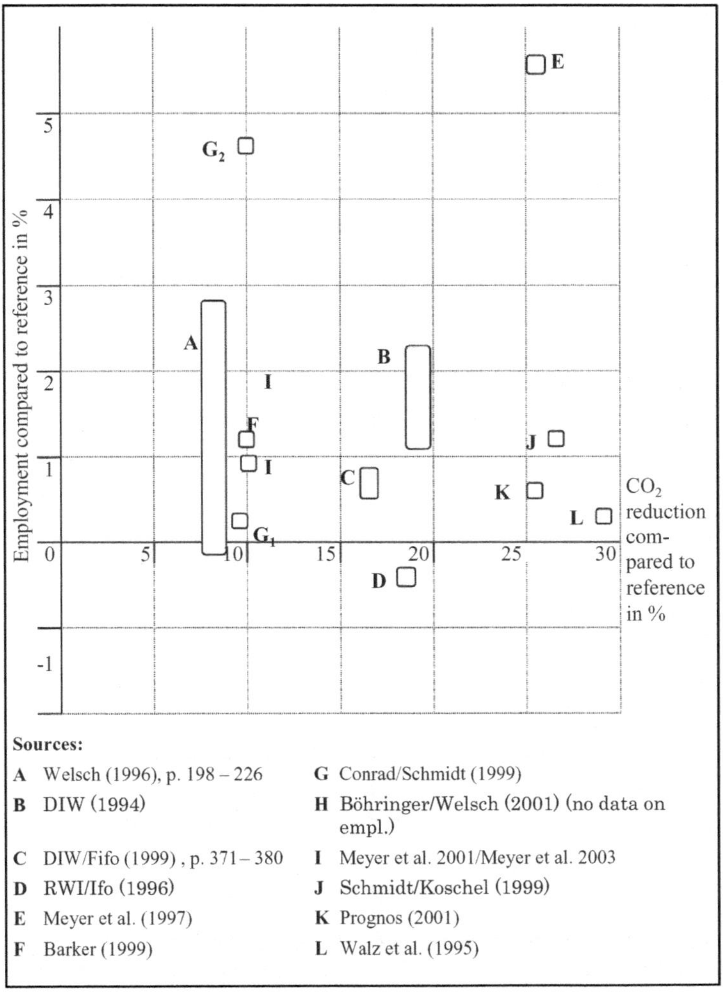

Fig. 3.1 The impacts of a climate policy on employment determined in various studies [Welsch, 1996, pp. 198–226; DIW, 1994; DIW/Fifo, 1999, pp. 371–380; RWI/Ifo, 1996; Meyer et al., 1997; Barker, 1999; Conrad and Schmidt, 1999; Böhringer et al., 2001 (no data on empl.); Meyer et al., 2001, 2003; Schmidt and Koschel, 1999; Prognos, 2001; Walz et al., 1995]

- In Welsch (upper range of the results), Conrad/Schmidt and Schmidt/Koschel, additional costs have an effect. These costs are implied by the GCE models, which assume an equilibrium without any no-regret potential as the starting situation. On the other hand, the changes in the distortionary effects of the tax systems and the labour cost reductions are effective.
- Böhringer et al. and Welsch (lower range of results) assume a direct transfer of the additional tax revenue to households in their analysis and suppress a reduction of distortionary taxes. The driving effect mechanism in the model is therefore the additional cost of the CO_2 reduction.

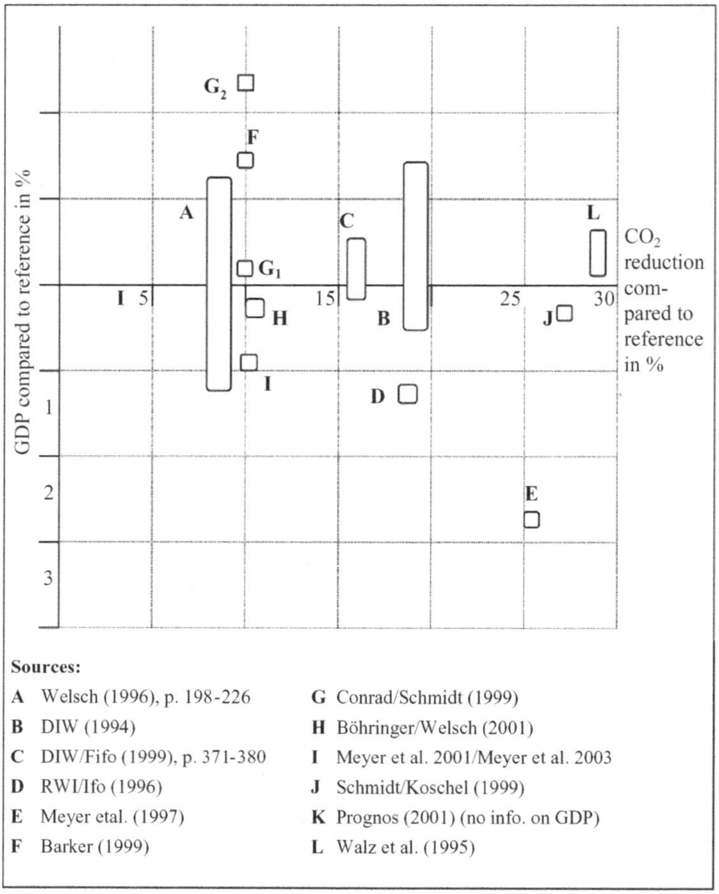

Fig. 3.2 The impacts of a climate policy on GDP determined in various surveys [Welsch, 1996, pp. 198–226; DIW, 1994; DIW/Fifo, 1999, pp. 371–380; RWI/Ifo, 1996; Meyer et al., 1997; Barker, 1999; Conrad and Schmidt, 1999; Böhringer et al., 2001; Meyer et al., 2001, 2003; Schmidt and Koschel, 1999; Prognos, 2001 (no info. on GDP); Walz et al., 1995]

- In the RWI/Ifo survey, aggregated top-down energy models are used combined with an econometric structural model. The existence of no-regret potentials is excluded in the reference scenario. At the same time, a comparatively short-term energy saving is simulated, which is not oriented on reinvestment cycles and is therefore extremely cost-intensive. The costs of these uneconomic saving measures are the decisive effect mechanism which results in negative impacts.[14]

[14] In sensitivity analyses with the same model, an energy tax used to reduce social security contributions results in positive employment effects. See Siebe (1996).

- In DIW (1994) and DIW/Fifo (1999), alongside the effects of introducing an energy tax and its compensation, the demand effects triggered by energy saving measures, which are depicted using the economic model applied, also play an important role in the positive impacts shown.
- In Meyer et al. (1997), the possible reduction in the cost of labour from redirecting energy tax revenue results in the production factor labour becoming relatively low-priced, thus substituting other production factors and there is a corresponding growth in its demand. Connected with this, there is a drop in productivity and a reduction of the GDP. Due to the high intensity of the measure examined, this simultaneously results in the highest growth in jobs and the largest GDP losses. A similar pattern, albeit, with a lower economic impulse, is found in Meyer et al. (2001, 2003), who used the same model.

In Barker (1999), who applies a sectorally disaggregated econometric model, the reduction in the tax burden of labour made possible by offsetting a CO_2 tax results in a greater demand for labour. The growth in macroeconomic production because of a strong double dividend is even more pronounced than the growth in employment, since productivity increases slightly. The reason behind this is an acceleration of technological progress which is endogenised in the E3ME model depending on the accumulated gross investments and the R&D expenditure. In Prognos, only the direct and indirect demand effects are modelled and correspondingly positive impacts estimated due to the application of a statistical input/output model. In contrast to all the other studies, a CO_2/energy tax is not examined specifically. With regard to the effect mechanisms considered, these results are to be regarded as problematic, methodologically, due to the limitations of the I/O analysis and the magnitude of the economic impulses analysed.

3.2.2 Results Using a Combined Bottom-Up/Top-Down Approach

In Sect. 3.3.1 the potential advantages were listed of linking the results of energy system analyses with a macroeconomic model. In contrast to a purely top-down method, different models are linked here which makes it much more complex. In general, such a combination could – and should – be implemented for both macroeconometric and CGE models. This section describes the method and results of two studies: First, the study from Prognos (2001), which analysed the effects of the German government programme to reduce CO_2-emissions, and second, the study for the Enquete Commission "Protecting the Earth's Atmosphere", which analysed the economic effects of reducing the CO_2 emissions by 40% up to 2020 using a combined bottom-up/top-down approach.[15]

[15] This work was commissioned by the German Bundestag. See Walz (1995a), Walz et al. (1995), and Walz (1997).

The study of Prognos (2001) used the detailed scenario results which were performed for the German government as an input. These scenarios were constructed with the help of various energy system models, among them the IKARUS model, and contain the effects on CO_2-emissions of various measures. The modelling approach uses an input-output model. Labour coefficients are forecasted for the future. Some technical change is introduced by changing the coefficients in the production inputs for investments needed. Three types of data form the key input into the model:[16]

- additional demand arising from investments into climate protection technologies,
- diminished demand for energy supply, and
- changes in consumption patterns, due to changing budget constraints of the households.

Thus, the approach mainly accounts for structural demand changes. Changes in costs are modelled implicitly, as long as they lead to changes in consumption demand. However, no effect of changing international competitiveness is accounted for. Furthermore, neither the effects of the climate policy on innovation nor a first mover advantage are modelled. Finally, the approach is not able to account for substitution effects of changing relative prices between the production inputs labour, capital and energy. Thus, effects such as the introduction of an energy tax with lowered labour taxation cannot be taken into account. The same problem arises for multiplier and accelerator effects, which cannot be modelled with a standard Input/Output model. Thus, to sum up, the study represents the typical weaknesses of the input/output models which were described in Sect. 3.1.

The study for the Enquete Commission also used energy scenarios based on the *IKARUS model* as starting-point of the analysis. Alongside a *reference scenario*, which assumes a constant nuclear capacity, a *reduction scenario* was calculated which assumes a nuclear phase-out at the same time as the CO_2 reduction of 40% compared with 1990. In addition, the entire spectrum of the measures discussed in the Enquete Commission forms the instrumental background for the reduction scenario. These include a moderate energy tax, which allows exceptions for the export-oriented and energy-intensive industrial branches. Additional measures focus on reducing the obstacles for an efficient use of energy, increasing the use of cogeneration and renewals, and increasing the energy efficiency standards. Simultaneously, limits are set, such as a minimum amount of domestic coal use, which restrict the substitution possibilities primarily in the electricity sector.

The *economic impulses* arising directly from these energy scenarios, e.g. investments and changes in energy costs and the energy tax yield compared to the reference scenario, form the input to a macroeconometric model and undergo further processing there. In contrast to the study of Prognos, however, a macroeconomically-oriented econometric long-term model of the DIW was applied.[17] The results

[16] Prognos (2001, pp. 67–76).

[17] See Blazejczak (1987), Chap. 3. Furthermore, a dynamic input/output model was used to analyse sectorial effects.

of the economic simulation calculations are deviations in the macroeconomic variables employment, GDP, inflation and private consumption induced by the reduction scenario compared to the reference scenario.

The model is open to embed the results of detailed, technology-based scenarios in determining the long-term macroeconomic income effects. The emphasis in the bottom-up analysis on a high degree of technological detail and clear possibilities of substituting energy by capital is expressed in additional investments and higher capital costs as well as in reduced energy costs and imports. The cost aspects are realised in the ratio of the capital costs to the saved energy costs; a double dividend due to the energy tax revenue and its recycling.

Compared to the reference scenario, the reduction scenario is linked with additional investments. Over the entire period under review the investments amount to around 350 billion Euros in 2000 prices. The magnitude of the *impulses* resulting from the reduction scenario is generally relatively small compared to the level of the macroeconomic aggregates in the simulation period.

The consequences of CO_2 abatement measures also depend on the economic framework. In general, it was assumed that the impulses are too small to justify fundamental changes of the behavioural assumptions. Different assumptions relating to the appearance of a double dividend, the significance of demand effects and the productive effect of climate protection technologies were made and bundled into two economic variants which group favourable and unfavourable conditions respectively. The two economic variants which result from this bundling reflect some of the main differences discussed in Chap. 2 regarding a pessimistic and an optimistic view of the effects of a climate policy (Table 3.1).

In the model simulations, there are only modest changes in the macroeconomic variables (Figs. 3.3 and 3.4). In the reduction scenario, the GNP is 0.3% or 0.7% higher in real terms on average over the period reviewed than in the reference scenario, depending on the variant. Employment also increases in each variant by ~0.3%. The main differences between the two variants are in *real private consumption*. In the unfavourable case, the CO_2 abatement measures are financed by a small reduction in the future increases of consumption, in the favourable case by increases in productivity. However, the small changes in private consumption have to be interpreted against the background of the development over time which leads to a

Table 3.1 Characterisation of the variants in "unfavourable" and "favourable conditions"

Effect mechanism	Variable	Unfavourable conditions	Favourable conditions
Double dividend	Use of energy tax revenue	Some consolidation of government budget	Tax reduction of indirect taxes
Demand effects	Housing investments	Some displacement of other consumer expenditure	Financing from loans and savings
Innovation effects	Companies' climate protection investments	Displacement of productive investments	Productive effect, modernisation of the national economy

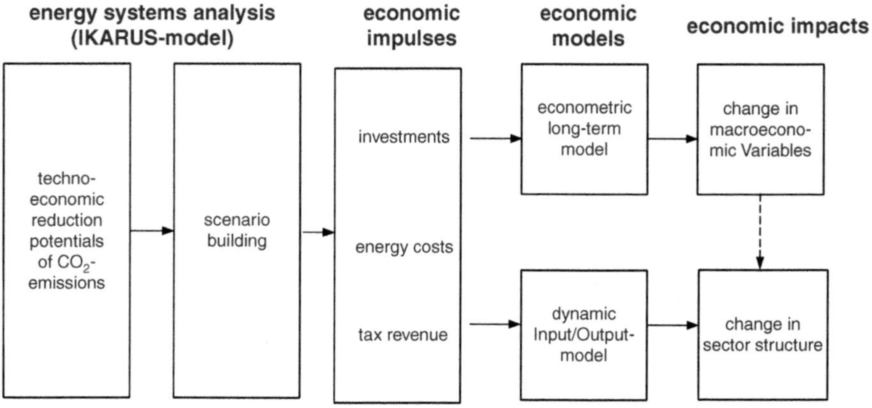

Fig. 3.3 Linking the models in the modelling approach

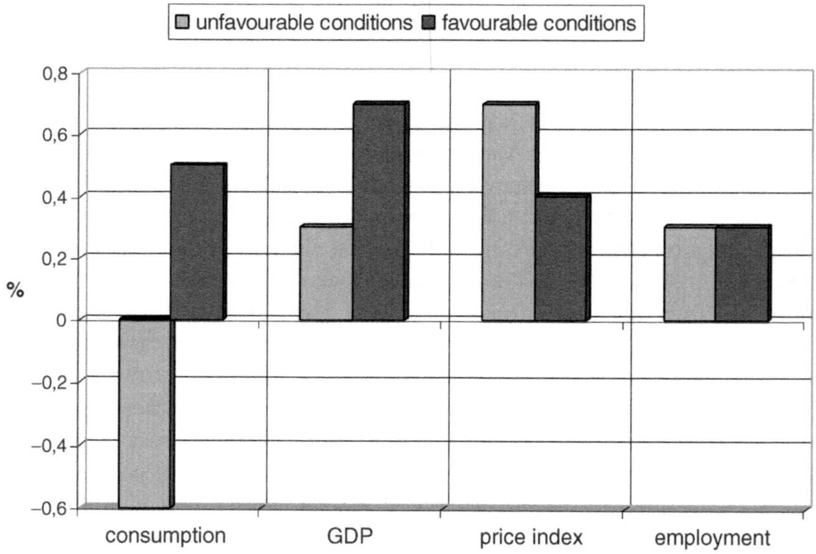

Fig. 3.4 Size of the macroeconomic impacts of the 40% reduction scenario with nuclear phase-out compared with the reference scenario in per cent (Walz et al., 1995)

doubling of the GNP and the variables derived from this between 1990 and 2020. In the most unfavourable case, therefore, the CO_2 abatement measures will not be financed by a reduction of the present consumption, but by doing without additional consumption to a small extent.

By setting up the two economic variants with favourable and unfavourable conditions, it is possible to measure off ranges to include effect mechanisms to a varying extent. For example, the crowding out of productive investments and consumer expenditure assumed under unfavourable conditions characterises a less favourable starting situation for demand side policy. In the favourable variant, the assumption of a productive effect of climate protection investments together with the modernisation of the capital stock makes sure that the effects on GDP are more positive. For this reason it is ultimately possible in this variant to finance climate protection investments macroeconomically without necessarily reducing private consumption. Because of the increased productivity, however, there is no corresponding increase in employment. Nevertheless, taken together, the results imply that the effects of a climate policy are rather modest.

Despite the broad range of likely development, which is covered by the favourable and unfavourable variant, many uncertainties remain:

- Firstly, the risks in foreign trade which are linked with a climate policy, e.g. for price competitiveness, are not fully explored.
- Secondly, the scenarios on the use of the revenue from an energy tax concentrate on lowering the budget deficit or indirect taxes. Thus, the effects of lower labour costs are not fully explored.
- Thirdly, the qualitative competitiveness in the sense of a first mover advantage is not included.
- Fourthly, this approach, despite its links with the IKARUS model, cannot sufficiently take into account the possible cost reductions resulting from policy-induced technological progress.

Due to the bottom-up foundation, a *substitution of energy by capital* is emphasised. In addition, the energy tax revenue is used to reduce indirect taxes but not directly reduce the taxes on labour. Overall, therefore, the effect on employment is likely to be relatively weak in comparison to other studies which emphasise the substitution effects in favour of labour. For a solid assessment of the economic impacts it is therefore necessary to put the results gained into context using the results of other studies in order to be able to make an evaluation of the impacts which reflects the entire range of results.

3.2.3 Comparison and Interpretation

A systematic comparison of the results is also made more difficult because there are many other factors involved apart from the models used and the correlations thus formed. For example, the studies refer to different reduction levels. These are not only due to the respective model, but also reflect a different intensity of the measures analysed. In addition, there are some distinct differences regarding the policy instruments analysed. Finally, differences in the simulation period also have to be taken into account.

Nevertheless, some *patterns* can be derived from the results. A reduction in the costs for labour, financed by the introduction of an energy tax, tends to cause positive job effects because of the substitution effects triggered between the production factors. Accordingly, the majority of studies show a rise in employment. At the same time, in most studies, the labour productivity falls slightly so that the impacts on production are usually much weaker.

The GCE models concentrate on price and cost effects. The results indicate that a strong double dividend seems possible for Germany, at least up to a reduction of ~10% of the CO_2 emissions. The isolated effect of removing distortionary taxes is also revealed in the differences in the results between the lower and upper variants of Welsch and Conrad/Schmidt. Due to the reduction of labour taxes, there is an increase of GDP and a strong double dividend.

Among the econometric models, the aggregated economic model of the DIW has comparably positive results. This model follows a Keynesian approach, in which the (net) demand effects play an important role. The rise in employment results here not only from the changed price relations of the production factors, but also from an increase in macroeconomic production which induces demand. On the other hand, it is obvious that the sectorally disaggregated econometric models set a comparatively low price elasticity of the energy demand. This implies relatively low energy savings or high tax rates and a high demand for adjustment. In accordance with this, Meyer et al. (1997) show a similar pattern of results as Schmidt and Koschel (1999), but the variations are much higher.[18]

The enormous *range* of the macroeconomic impacts within one single study is also worth noting.[19] The variations in the macroeconomic framework conditions of energy saving are responsible for this. For example, a CO_2/energy tax – introduced for climate policy reasons – can result in quite different macroeconomic effects depending on whether the tax revenue is used to reduce the taxes on labour, to decrease the national debt, or to increase public spending. In the DIW work, in addition, there are also different assumptions with regard to the development of the exchange rate, monetary policy and the substitution effects. On top of this, both the assumed flexibility of the capital stocks and assumptions about the labour market play an important role in the variations within the individual studies.[20] The magnitude of these differences within individual studies makes it very clear that not only the actual energy policy is of significance for the macroeconomic impacts, but also, to a considerable extent, how it is embedded in the economic policy.

When comparing the results of the combined top-down/bottom up approach with top-down studies, the most obvious difference is that the economic impacts in

---

[18] For example, the elasticities in Koschel/Schmidt are approx. twice as high as in Meyer et al. In spite of the same review period in both surveys, differences in the short-term and long-term elasticity may be reflected here. Meyer et al. with their econometric model depict rather short-term elasticities, whereas the substitution processes in an equilibrium model correspond logically to end points of the effect and therefore long-term elasticities.

[19] This is particularly true for Welsch (1996), DIW (1994) and Conrad and Schmidt (1999).

[20] For example, in DIW/Fifo (1999), two variants are distinguished which differ regarding the assumptions made about wage policy. In the case of a moderate wage policy, there are more favourable macroeconomic effects than in the case of an aggressive one. In a similar way, modelling the job market with a considerable reduction of real wages (Conrad and Schmidt, 1999, Variant G_2) results in much higher GDP, increased employment and reduced private consumption.

the combined approach are relatively weak in spite of high CO_2 reduction. This is firstly because the results are based on the year 2020, whereas most other studies have a much shorter *time horizon*, with the exception of Welsch and Prognos. This means that the more intensive adjustments are able to be extended over a longer period of time.[21] Secondly, the different *foundation* of the *input data* must be considered: the impulses triggered by the reduction of emissions are based on technically detailed and up-to-date results of technology forecasting, whereas in the econometric top-down models, elasticity estimates are used based on the past and tending to refer to the structures of the 1970s and 1980s.[22] A third reason for the comparatively small effect is that mechanisms are taken into account which counteract the negative effects of the increasing marginal costs of CO_2 reduction. These mechanisms include the demand effects and the productive effects of the industrial climate protection technologies which are beginning to be considered in the favourable variant.

In addition, comparing the following patterns of results of individual studies can provide interesting indications for understanding the *similarities* and *differences*:

- The work of Meyer et al. and Schmidt/Koschel arrive at similarly high reductions compared to the reference case, although these are based on a much shorter adjustment period. Unlike these two studies, the combined bottom-up/top-down approach emphasises the substitution of energy by capital. As a result, employment growth is considerably smaller, but there is no reduction of productivity and GDP.
- Welsch discusses an energy tax as a single measure. In a similar form, this tax is part of the catalogue of measures analysed with the combined bottom-up/top down approach. Welsch shows that the cost burden of the energy-intensive sectors originating from this kind of energy tax does not involve any significant competitiveness problems.[23] These results indicate that the foreign trade risks mentioned in Sect. 3.2.2 are not likely to be very pronounced.
- The unfavourable variant of the bottom-up/top-down approach runs parallel to Meyer et al. and Böhringer et al. to the extent that a clear increase in investments is financed in each study by a slight reduction in private consumption.[24] Similar to the favourable variant, both investments and private consumption increase in Welsch (upper variant), DIW/FiFo and Barker.[25] Looking at this in more detail reveals differences which can be attributed to the different substitution strategies: for example, in the latter studies, private consumption increases to a much

[21] As in the majority of analyses listed, the annual emission reduction is in the order of 1% per year compared to the reference scenario. RWI/Ifo (1996), Meyer et al. (1997) and Schmidt and Koschel (1999) imply much higher annual adjustment demands.

[22] A specific problem of these top-down analyses is the gap in the statistical time series due to the German reunification which complicate data updating.

[23] This is particularly true for the sensitivity analysis in which Welsch examines the effect of much higher substitution elasticity between domestic and foreign production than he assumed in the base variant. See Welsch (1996, pp. 216–219).

[24] See Meyer et al. (1997, p. 20), Böhringer et al. (2001, p. 9).

[25] See Welsch (1996, p. 211), DIW/Fifo (1999, p. 375), Barker (1999, p. 413).

greater extent than investments; at the same time, there is a much stronger increase in employment. In contrast, investments increase considerably in the combined top down/bottom up approach. This expresses the substitution of energy by capital, whereas employment rises less strongly in comparison to the other studies.

Overall, the conclusion can be drawn that the results obtained using the combined bottom-up/top-down approach move within a comparable framework as the other studies, especially if the longer simulation period is considered. A special characteristic which stands out in comparison to the other studies is the emphasis on the substitution of energy by capital caused by the bottom-up foundation. A second difference is that the change in employment is not relatively more positive than the change in production, because the additional tax revenues are not used to reduce the taxes on labour. Finally, the investments in climate protection in the favourable variant also have a productivity-increasing effect. Under favourable conditions, therefore, the GDP grows even more than employment – a characteristic which is only found in Barker's work modelling an endogenisation of technological progress.

Ultimately the difference in the results reflects the different strategy of the climate policy. If there is a mix of instruments, and the height of the CO_2/energy tax is limited and a considerable share of investments in climate protection is steered towards the space heating sector, results as discussed in Sect. 3.2.2 are plausible. If, in contrast, a uniform and very high CO_2/energy tax plays a dominating role and there are no special regulations for the energy-intensive sectors, the effects on production will be less positive. However, the detrimental effects on employment will be overcompensated by the reduction of labour costs leading to overall positive employment effects. With a moderate energy tax, such as that introduced in Germany, which is indeed used to reduce social security contributions, but where the main burden of the energy price increases is directed towards the households and service sector and where the production sector in general and the energy-intensive sectors in particular are given preferential treatment under special regulations,[26] it should be expected that the results will lie somewhere in-between.

When interpreting the studies shown in Figs. 3.1 and 3.2 it must be considered that they refer to the supply or demand effects of a climate protection policy, but that the *innovation effects* triggered by this are not or not sufficiently taken into account. The technological change induced by the climate policy is not taken into account other than conceivable – but not easily quantifiable – first mover advantages. The positive productivity effects of climate protection technologies which may possibly occur are only considered in the form of a variation calculation in the combined bottom-up/top-down approach and in Barker. All in all, it is therefore reasonable to assume that the model analyses cited above would arrive at more positive impacts on the economy as a whole if they took these innovation effects into account as well. The deficiencies of the macroeconomic models regarding the

[26] See Linscheidt and Truger (2000).

adequate consideration of technological progress could result in a tendency to distort the results. This underlines the necessity to direct future research towards a better modelling of technological progress.

Overall, the following *conclusions* can be drawn with regard to the macroeconomic impacts of alternative developments:

- A combined bottom-up/top-down approach stresses the substitution of energy by capital, whereas the results derived from abstract production functions also show clear substitutions of energy by labour.
- The top-down approaches predominantly model ecological tax reforms, whose revenue is used to lower other distortionary taxes, mostly on labour. This reinforces the employment effects shown in comparison to modelling a package of measures in which considerable CO_2 reductions are targeted in the space heating sector. The differences in the results can thus be explained by the fact that different energy policy strategies are being analysed.
- Depending on the assumption how the energy policy is embedded within the overall economy (e.g. form of tax recycling, assumptions about productivity, behaviour of economic agents), the results differ substantially already *within* the studies. This indicates that differences with regard to the stringency of targets and the instruments of the energy policy are perhaps less decisive for the macroeconomic effects than how this policy is embedded within the economic policy.
- The survey for Germany generally shows moderate macroeconomic effects. According to the results, it is plausible that a short-term moderate reduction in CO_2 emissions in the order of 10% would indeed result in an increase in employment and probably GDP, too.
- In the medium to long term, a reduction of the CO_2 emissions in the order of 40% will only marginally change the GDP and bring about a moderate increase in employment.[27] In total, a growth in employment from 200,000 to 300,000 seems feasible. Whereas, under favourable conditions, a minimum increase in private consumption is conceivable, under less favourable conditions there are likely to be slight reductions in private consumption.
- Combining bottom-up and top-down methods provides a good starting-point for establishing micro-macro bridges within technology-economic analyses and should be refined methodologically in the future and applied more often.
- The positive effects could be amplified even more by the innovation effects, which are not sufficiently considered in the model results.

---

[27] This kind of result is found between the results calculated with the bottom-up/top-down approach and those of Schmidt and Koschel (1999) using the ZEW model. It is plausible because an energy tax is likely to be used at least to some extent to reduce the tax burden on labour unlike the policy assumptions made using the bottom-up/top-down approach. The substitution effects resulting from the reduced price of labour compared to all other production factors cause an additional increase in employment. However, this effect will probably not be too pronounced because a climate policy does not just use price instruments and thus higher energy tax rates and corresponding reductions of social security contribution, but applies a mix of instruments instead.

Chapter 4
Structural Adjustments

4.1 Scope of Analysis

Chapter 3 concluded that a major change in the overall German production is not a likely outcome of the macroeconomic analysis. With regard to employment, the most probable result is a modest increase in the number of jobs. However, in addition to the macroeconomic effects, the structural adjustments faced by the economy are of interest, too. There are three different structural effects which are analysed in this chapter:

- sectoral effects resulting from the shift between economic branches,
- regional effects showing shifts in employment between the different regions, and
- structural effects with regard to job characteristics and qualification requirements.

The sectoral effects form the starting point of the analysis. A review was made of the results of macroeconomic studies which include a sectoral analysis. A more detailed examination is performed for two scenarios with a disaggregated sectoral analysis. The effects of the German Ecotax were analysed using an econometric model based on input-output tables (Ecotax scenario) within the studies from Meyer et al. (2001) in combination with Meyer et al. (2003). The study of Walz et al. (1995) analysed the effects of a policy mix to reduce CO_2 emissions (policy mix scenario). Within that study, the sectoral changes were analysed using a dynamic input-output model which had been linked to an aggregated macroeconometric model.

These sectoral changes also imply different effects on the economic activities within the regions. Depending on the role the specific economic sectors play in the respective region, some regions might experience an above average increase in employment, while others might lose jobs despite an overall growth in employment nationwide. Thus, the section on regional employment identifies the likely effects of the Ecotax scenario and the policy mix scenario here. For both scenarios, the **I**ntegrated **S**ustainabil**i**ty Assessment **S**ystem (ISIS), which disaggregates

R. Walz and J. Schleich. *The Economics of Climate Change Policies.*
Sustainability and Innovation,

the German economy into 181 regions,[1] is used to address the following questions with different indicators:

- changes in employment per labour district, and distribution of relative winners and losers,
- relative gross changes in employment (labour turnovers), indicating the demands on the functioning of regional labour markets,
- changes in East Germany versus West Germany, and
- change in overall regional concentration of economic activities using a Herfindahl Index of regional concentration.

The sectoral changes also imply different effects on the job characteristics and qualification requirements. In order to account for these structural adjustments, the ISIS model was used again. It contains a sub-module which describes the job characteristics and qualification requirements within each of the economic sectors. Thus, the two scenarios are also analysed with regard to following indicators:

- qualification requirements (master degree, bachelor degree, foreman/technician, apprenticeship, without education/training),
- percentage of part-time jobs and jobs with a limited term job contract, and
- percentage of jobs with an increased need for flexible working hours (weekend/holiday work; evening/night work; shift work).

4.2 Sectoral Changes

Not all of the studies listed in Chap. 3 show sectoral impacts as well. Furthermore, most of the sectoral results are only available at a rather aggregated level. However, it is clear from these results that, as expected, energy suppliers are the main losers of climate protection policies. The employment-intensive service sector is one of the relatively favoured sectors in the analyses which model the effects of a CO_2 tax used to reduce the tax burden on labour.[2] At the same time, the energy-intensive basic industries are more strongly affected by this kind of policy so that they have to bear some disproportionate losses. The assessment of the impacts on the capital goods sectors and the building industry differ by author.[3] In Welsch (1996), the building industry and the capital goods industries do better than average. Here, relatively small energy cost ratios and higher labour cost ratios overlap the changes in

[1] See Annex for a detailed description of the ISIS model.

[2] See Schmidt and Koschel (1999, p. 166–167); Barker (1999, p. 409); Welsch (1996, p. 212); Meyer et al. (1997, p. 14–15).

[3] Whereas both industrial branches are only averagely affected in Schmidt and Koschel (1999, p. 166–167) and Barker (1999, p. 409), in Meyer et al. (1997, p. 14–15), they are among the relatively favoured ones. In Welsch 1996, the building industries and the capital goods industries do better than average.

the impacts for the macroeconomic aggregates investment and exports, on which these sectors are disproportionately dependent.

Some interesting results on the sectoral effects of a very high CO_2 tax can be drawn from Meyer et al. (1997). The assumed CO_2 tax differentiates between the energy sources according to their carbon content. It implies severe changes in the energy producing sectors, leading to a very marked reduction in coal use and hence coal production. Furthermore, a CO_2 tax favours natural gas over mineral oil. Thus, there are substitutions away from oil. This results in coal and oil being the main losers among the energy producing sectors. At the same time, the high CO_2 tax permits more than just modest reductions in social security contributions. Labour-intensive sectors benefit greatly from this policy. Thus, these sectors tend to be favoured in this scenario. As service-oriented sectors generally have a rather high labour intensity, they also tend to make above average gains. To sum up, the very high CO_2 tax scenario implies severe structural changes within the energy supply sectors and between energy-intensive production and service-oriented and more labour-intensive businesses.

For a more exact analysis of the structural impacts it is necessary to take a more disaggregated approach. Furthermore, it is important to analyse policy approaches which emphasise the political feasibility of CO_2 reduction policies. The following macroeconomic studies fulfil these conditions. First, the study by Meyer et al. (2001) in combination with Meyer et al. (2003): in this analysis, the scenario of an extended German Ecotax was modelled using the Panta Rhei model. Second, the study by Walz et al. (1995), which analysed a policy mix scenario of the Enquete-Commission of the German Bundestag to reduce CO_2 emissions. Within this study, the sectoral changes were analysed in a combined top-down/bottom-up approach using a dynamic input-output model linked to an aggregated macroeconometric model. Both studies have the advantage that they account for additional economic mechanisms which a traditional sectoral analysis, based only on demand shifts and a partial model, must neglect. Insofar they paint a more detailed picture of the sectoral effects of a CO_2 reduction policy.

It is important to keep in mind that the overall stringency of the measures varies: in the Ecotax scenario, on the one hand, with a more short-run but less intensive policy impulse; and in the policy mix scenario, on the other, with a medium-term policy impulse, which is, however, not just price-triggered. Furthermore, the results were obtained by a different modelling approach, with the Ecotax scenario highlighting the substitution of energy by labour, and the policy mix scenario emphasising the substitution of energy by capital.[4] Thus, the differences can be compared only with regard to the pattern of sectoral effects, not with regard to the overall change within each sector.

The changes in sectoral output are reproduced in Fig. 4.1. Figure 4.2 shows the change in the share of the sector at total production and employment, respectively.

[4] See Chap. 3. Thus, the Ecotax scenario shows rather positive employment and negative production effects.

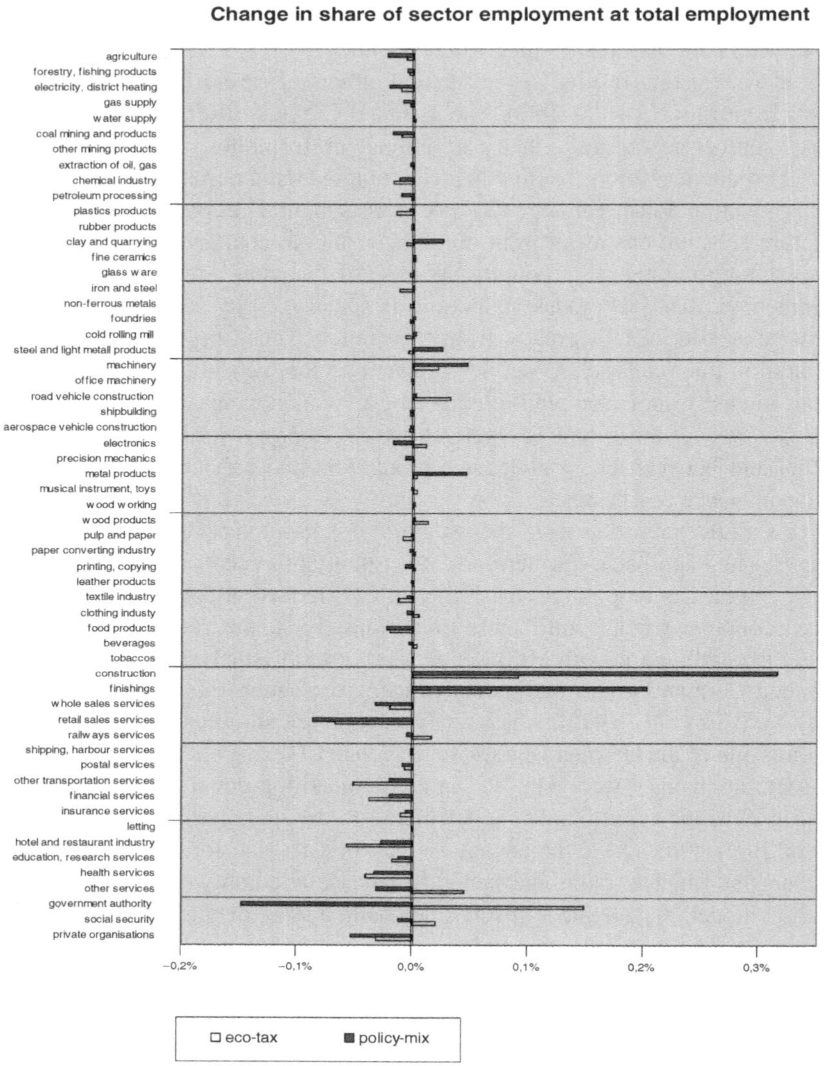

Fig. 4.1 Changes in production in the Ecotax and the policy mix scenario in %

This makes it easier to identify a changing pattern of relative importance of the sector; because sectors which become relatively more important have a positive sign (winners) and those losing relatively have a negative sign (losers).

In both scenarios, the energy supply sectors are losers. However, there are differences between the scenarios. In the Ecotax scenario, the electricity sector loses relatively more production than in the policy mix scenario. On the other hand, gas loses more in the policy mix scenario. There are also some similarities with regard to the winners. In both scenarios, construction and associated industries closely

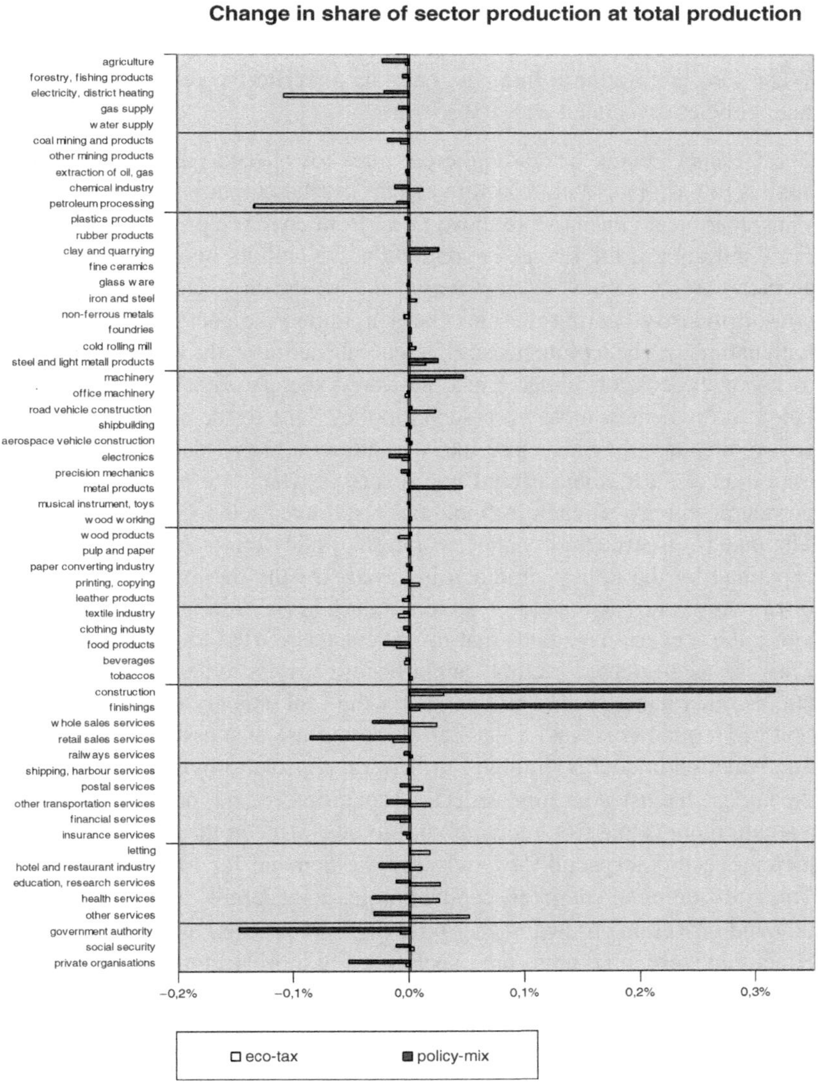

Fig. 4.2 Changes in shares of sector in total production in the Ecotax and the policy mix scenarios in percentage points

related with regard to production technology (such as minerals or steel and light metal construction) are winners. The same holds for some capital goods sectors, especially machinery, and production-related sectors such as metal working. However, there is one important differentiation. Within the policy mix scenario, it is predominantly the machinery sector among the capital goods which increases production. In the Ecotax scenario, the relative gain of this sector is lower, but there are a few other capital goods sectors which benefit considerably. In addition, there are also a few service sectors plus the government sector among the winners, which all lose importance in the policy mix scenario.

When interpreting these findings, one has to relate the results to the different climate policies associated with the scenarios:

- The German Ecotax increases the tax rates for oil and gas. A special tax rate applies to gasoline. With regard to electricity, the output is taxed, not the input. Thus, there is no incentive to move away from coal as a primary energy source. For the design of the Ecotax beyond 2004, the authors of the scenario assumed an increase in tax rates for electricity and gasoline only. This results in the mineral oil industry bearing the main burden among the energy supply industries, with natural gas being much less affected. Furthermore, the tax revenues are used to lower the cost of labour. Thus, industrial sectors with low energy but high labour costs benefit most from such a policy. Due to the above average labour cost reductions, they are more likely to attract a higher demand and to gain in importance. Thus, some capital goods sectors which are not closely related to producing energy-efficient technologies also increase their production.
- The pattern of structural change within the policy mix scenario can be largely explained by the policy approach involved. On the one hand, different policy instruments were assumed to lead to a reduced role of tax instruments. Furthermore, the scenario presumes that tax revenues are used to lower indirect taxes. Thus, service-oriented sectors cannot benefit from a reduction of labour costs. On the other hand, special provisions for the coal industry (a minimum production of German coal) and a nuclear phase-out are also assumed, implying that the reduction in coal is limited. Furthermore, special provisions were made for the energy-intensive sectors, reflecting conditions of the political economy. As a result, more of the CO_2 reduction has to take place in the household and commercial sectors, especially by reducing the demand for low temperature heat. This substitution of energy by capital implies considerable increases in construction and in sectors producing energy-saving investments. In general, this has the effect that substantial gains are experienced in those sectors with close production ties to the construction industry or the investment industries. This also reflects the modelling approach chosen within this scenario, which highlights the substitution of energy by capital.
- Chapter 3 concluded that the employment effects of a CO_2 reduction policy will be slightly more positive than the rather negligible effects on GDP. This result can also be seen in the two studies with detailed sectoral disaggregation. Figures 4.3 and 4.4 depict these results for the Ecotax and the policy mix scenarios by showing how the employment in each sector is affected and how the share of each sector in total employment changes.

Even under the assumption of a zero effect on GDP, the study of Walz et al. (1995) shows an increase in employment of about 0.2%. Without using tax revenues to lower the cost of labour, the study assumes that the productivity in each sector remains largely unchanged by substitution effects between the production factors. Thus, the increase in employment reflects only the shift towards more

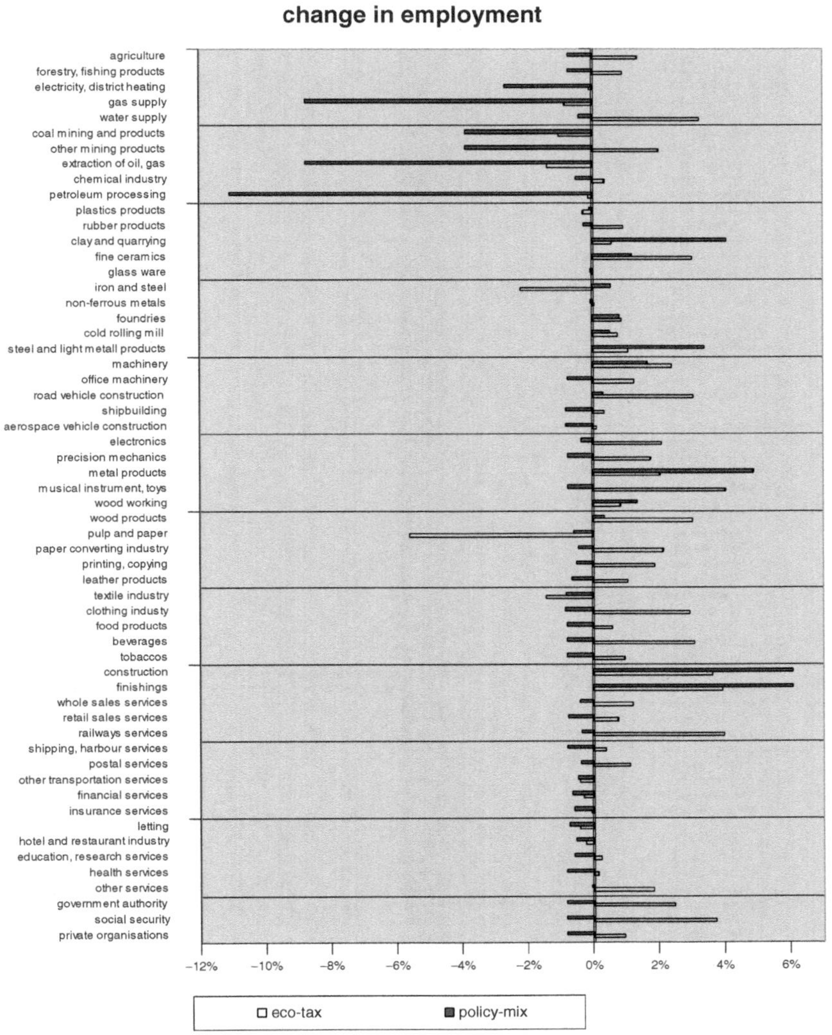

Fig. 4.3 Changes in sector employment in the Ecotax and the policy mix scenarios in %

labour-intensive sectors. The study of Meyer et al. (2001, 2003) presents an even larger growth in employment of 1.4%. This effect is not only due to the shift towards more labour-intensive sectors, but is mostly driven by labour becoming cheaper due to the recycling of the tax revenues. Thus, the reduction in labour costs brings about an overall increase in labour per production unit. In general, this leads to changes in sectoral employment being more pronounced than changes in production. Furthermore, this effect helps to explain the above average increase in employment for some of the capital goods sectors as well as for a few service sectors and government. However, as already mentioned in Chap. 3, the magnitude of this effect

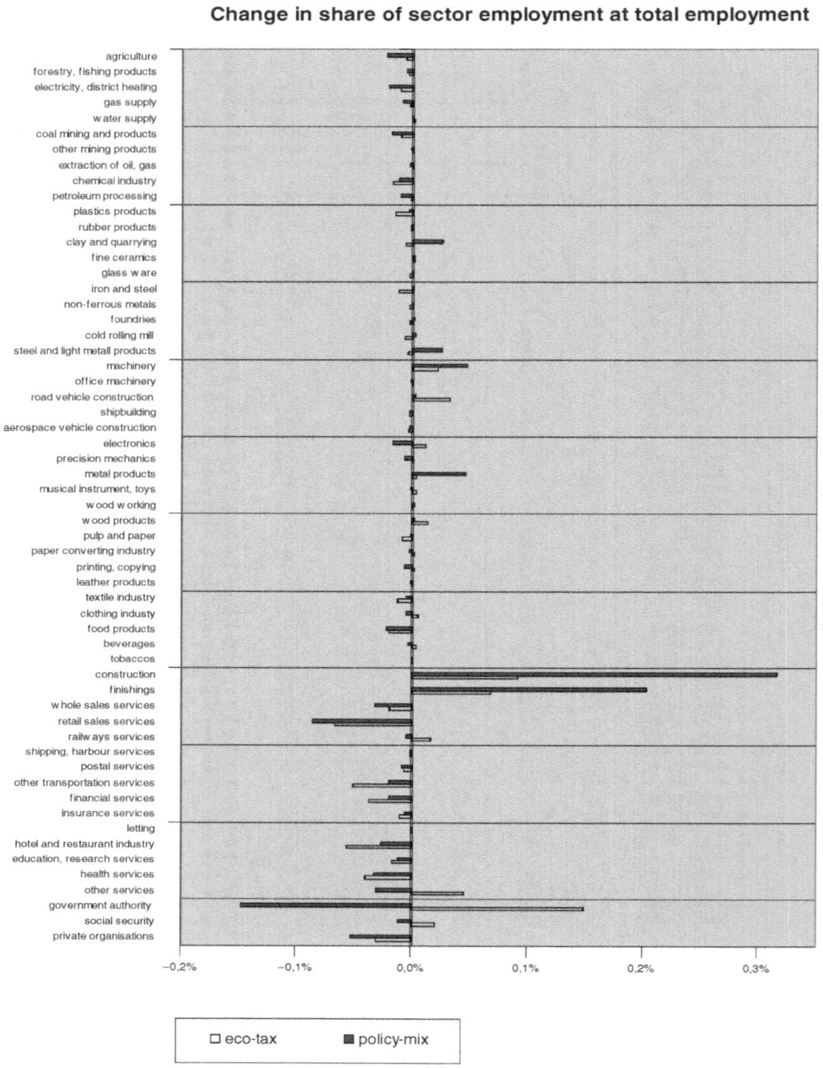

Fig. 4.4 Changes in shares of sector in total employment in the Ecotax scenario and the policy scenario compared to the reference case in percentage points

also depends on the modelling approach chosen. It has to be kept in mind that the model used for the Ecotax scenario is among those producing the biggest discrepancies between production and (positive) employment effects. Thus, the results shown reflect the upper end of the estimates about the effect of employment increases due to the substitution effects of labour becoming relatively less expensive.

To sum up the sectoral results, it can be assumed that the energy supply industry will be a loser under a climate protection policy, unless it engages in CO_2-reducing technology. The differences which can be found between the different studies mainly reflect varying assumptions about policy details, which, to a certain extent, can be traced back to the underlying political economy of a climate policy. This is especially true for the future role of coal, and the possible future of natural gas. Whether or not the service sectors are among the winners also largely depends on policy details. If, for example, the climate policy mainly relies on a uniform CO_2 tax, the revenue from which is used to lower taxes on labour, they are likely to benefit. If, however, a tax policy is implemented which does not affect energy use according to its carbon emissions, but puts greater emphasis on increasing prices for gasoline and electricity, the service sector is less likely to benefit. Moreover, if the main burden of reduction is carried by households and small commercial sectors[5] and the instruments used do not lead to tax revenues lowering the costs of labour, the service sector is likely not to benefit at all. The construction-related sectors will be among the clear winners of a CO_2 reduction policy, together with some of the capital goods industries.

4.3 Changes in Regional Employment

4.3.1 Scope of Analysis and Methodological Approach

The sectoral changes of a climate policy also imply different effects on the economic activities within regions. Depending on the role the specific economic sectors play in the respective region, some regions might experience an above average increase in employment, while others might lose jobs despite an overall growth in employment nationwide. Thus, in the section on regional employment, the likely effects of CO_2 reduction policies on regional employment are identified. However, this requires the use of a model which not only distinguishes sectoral employment, but is also able to disaggregate sectoral employment even further on a regional level.

One such model is the Integrated Sustainability Assessment System (ISIS) developed at FhG-ISI (see the description in the Annex). It is based on the standard input-output tables of the German Federal Statistical Office which disaggregate the German economy into 58 sectors. In addition, this model has been augmented with a regional sub-module which disaggregates each sector still further into the 181 districts of the federal labour office. This model is used to calculate the regional effects of CO_2 reduction policies.

The analysis of the regional effects requires the sectoral changes as a data input. In this study, the following approach was taken. The results of the two macroeconomic studies with a detailed analysis of sectoral change were taken as data input.

[5] This is the case in most scenarios for German policymakers, such as in the policy mix scenario.

The translation of the model results into the ISIS model proved to be unproblematic, as all the models were constructed using a common classification of sectors. In order to reflect the sectoral change induced by the Ecotax or a policy mix, the sectoral changes in employment of the two mentioned studies compared to the reference case were used as data input.

The model runs of the ISIS model yield the relative change in regional distribution of economic activities. The regional disaggregation is performed at the level of the 181 German labour office districts, on which the ISIS regional sub-module is based. These results form the basis of the interpretation of regional effects with different indicators. The following questions are addressed:

- Changes in employment per labour district, and the distribution of relative winners and losers.
- Relative gross changes in employment (labour turnovers), indicating the demands on the functioning of regional labour markets.
- Changes in East Germany versus West Germany.
- Change in overall regional concentration of economic activities using a Herfindahl Index of regional concentration.

4.3.2 Net Changes in Employment in the Labour Office Districts

Even though both scenarios lead to an increase in jobs, there are also labour office districts which lose jobs. However, given the different sizes in labour office districts, the absolute number of jobs can be a misleading indicator for the severity of these changes. Thus, in order to account for differences in the overall size of the labour office districts, the relative changes (absolute changes in relation to the number of jobs in each district) were used as the indicator. Furthermore, the analysis concentrates on the relative winners and losers of the regions in order to emphasise the pattern of regional change. Therefore the deviations of the changes in employment from the arithmetic mean for all labour office districts are used as an indicator. This also has the effect that the results for the two policy scenarios are easier to compare.

Tables 4.1 and 4.2 indicate the deviation from the average change in jobs for both the top ten winners and top ten losers. Most of the clear losers in the Ecotax scenario (Table 4.1) are among the coal mining regions in Germany. Within those regions, the coal-related sectors are the ones which lose the most shares in total employment within the labour district. They are so important within these labour districts that their loss in employment results in a below average development of the total district. Two other labour districts (Duisburg and Ludwigshafen) have high shares of iron and steel and chemistry, respectively, which are characterized by a below average development in employment. In addition, an agglomeration is found in Frankfurt/M. with a lower share of those sectors which gain the most and a very high share of transportation, which is among the losing sectors. This fact is also responsible for the labour district Freising being among the losers. With two

Table 4.1 Relative net changes in employment in the Ecotax scenario

Top ten winners and losers in relative changes in employment (percentage points above/below average rate of change in jobs)

Top ten winners		Top ten losers	
Labour office district	% points	Labour office district	% points
Helmstedt	0.45	Duisburg	−0.41
Berlin East	0.42	Recklinghausen	−0.40
Landshut	0.32	Düren	−0.37
Magdeburg	0.29	Gelsenkirchen	−0.33
Nordhausen	0.28	Ludwigshafen	−0.29
Erfurt	0.27	Hamm	−0.29
Stendal	0.27	Frankfurt/M.	−0.23
Sangerhausen	0.25	Freising	−0.23
Zwickau	0.24	Rheine	−0.23
Dessau	0.24	Brühl	−0.23
Weighted average top ten	0.30	Weighted average top ten	−0.28

Table 4.2 Relative net changes in employment in the policy mix scenario

Biggest winners and losers in relative changes in employment (percentage points above/below average rate of change for total economy)

Top ten winners		Top ten losers	
Labour office district	% points	Labour office district	% points
Oschatz	0.62	Gelsenkirchen	−0.73
Riesa	0.61	Recklinghausen	−0.64
Pirna	0.61	Berlin Mitte	−0.42
Altenburg	0.55	Saarbrücken	−0.41
Zwickau	0.52	Münster	−0.4
Sangerhausen	0.52	Hamburg	−0.39
Nordhausen	0.50	Köln	−0.39
Bautzen	0.48	München	−0.35
Iserlohn	0.47	Frankfurt/M.	−0.34
Siegen	0.45	Koblenz	−0.32
Weighted average	0.52	Weighted average	−0.40

exceptions, the top winners are all located in East Germany. This reflects the high shares of construction-related sectors plus government in these labour market districts. The two West German labour market districts among the top ten winners are characterized by extremely high shares of the car industry, which experiences above average growth in employment under the Ecotax scenario.

Some of the clear losers in the policy mix scenario are also among the coal producing regions (Table 4.2). Even more than in the Ecotax scenario, there are some urban agglomerations among the losers. They are characterized by lower importance of the winning sectors (construction related, machinery) on the one hand, and a higher importance of service sectors and government on the other, which both

Table 4.3 Distribution of relative net wins and net losses for all labour office districts

	Ecotax scenario	Policy mix scenario
Number of districts above average	90	114
Number of districts below average	90	66
Deviation of median labour district from arithmetic average (percentage points)	–0.003	0.076

drop in significance in the policy mix scenario. The clear winners are mostly East German regions. They are characterized by an above average importance of construction on the one hand, plus a significant importance of machinery-related industries on the other. Thus, it is not surprising to find most of these winning East German regions located in Saxony, which has the highest share in capital good related production among the East German regions.

The weighted average for the top ten winners and losers implies only a minor difference in the distribution of wins and losses. For both the Ecotax and the policy mix scenario, the deviation of the top winners from the average does not differ greatly from the deviation of the top losers.

The analysis so far only covered the top ten winning and losing regions. A more thorough analysis must also take the distribution of wins and losses among all the regions into account. First indications of this distribution can be deducted from Table 4.3. In the Ecotax scenario, the number of regions below average equals the regions with above average changes in employment. The median being equal to the arithmetic mean indicates that the overall distribution of losses and wins is quite symmetric. The results for the policy mix scenario present a somewhat different picture. The regions above average clearly outnumber the regions below average. Thus, the deviation of the median region from the arithmetic mean is clearly positive. This relates to an uneven distribution of wins and losses. In general, the losses tend to be more pronounced than the gains in the respective sectors. However, this does not hold for every segment of the winning and losing sectors, as the results for the top ten winners and losers showed.

4.3.3 Gross Changes in Number of Jobs

The sectoral changes imply that there are both increases and decreases in the regions. However, the analysis in Sect. 4.2.1 concentrated on the net employment effects in the regions only. Thus, the positive effects and the negative effects cancel themselves out to a certain extent. However, both a loss and a gain of a job imply that the labour market must perform. If labour markets are sticky, it might be difficult to fill additional jobs with persons who lose their jobs. This holds for both job turnovers within one region, and movements of labour from one region to the other. Thus, in order to account for the adjustment pressure which CO_2 reduction policies impose on the regional labour market, the gross number of changes in jobs was calculated by adding the (positive) number of job losses to the number of job increases. The higher the percentage of gross job changes compared to total employment in the district (job turnover rate), the more severe is the additional effect of the CO_2 reduction policy on the functioning of the regional labour markets.

The analysis of all the labour districts for the Ecotax scenario yields a job turnover rate of 1.5% of all jobs. However, the small difference from the net change in jobs makes it clear that the overall number of jobs lost is not very significant in this scenario. Table 4.4 shows the job turnover rate for the regions with the lowest and the highest gross changes. It is not surprising to see that a number of the top relative winners also have the highest demand on the labour market. On the other hand, among the regions with the lowest turnover rates are some of the top losers. Both aspects underline that the labour market is mainly driven by the increase in jobs.

The analysis of all labour districts for the policy mix scenario yields a job turnover rate of 1.2% of all jobs (Table 4.5). As in the Ecotax scenario, many of the biggest winners can be found among the regions with the highest gross turnover rate. However, in contrast to the Ecotax scenario, the gross turnover rate exceeds the net wins by approximately a factor of three. Thus, the effect on the labour market is substantially greater than the net effects imply and cannot be neglected. Some of the losers can be found among the regions with the lowest turnover rate. This reflects the fact that these regions are mainly among the losers because they do not benefit much from the job increases in the growing sectors, and not because the number of job losses is extraordinarily high.

4.3.4 Comparison of East and West Germany

The distribution of employment changes among the regions is also of importance with regard to the economic development within East Germany. Ever since unification, the economic development in East Germany has been of special interest. Thus,

Table 4.4 Job turnover rates in the Ecotax scenario

Job turnovers in per cent of total jobs			
Lowest demand on labour market		Highest demand on labour market	
Labour office district	% points	Labour office district	% points
Ludwigshafen	1.20	Helmstedt	1.94
Frankfurt/M.	1.30	Berlin East	1.87
Goslar	1.31	Landshut	1.78
Hamburg	1.31	Nordhausen	1.77
Konstanz	1.32	Zwickau	1.75
Freising	1.33	Stendal	1.75
Mönchengladbach	1.33	Magdeburg	1.75
Recklinghausen	1.33	Erfurt	1.73
Gelsenkirchen	1.33	Chemnitz	1.73
Münster	1.34	Pirna	1.73

Table 4.5 Job turnover rates in the policy mix scenario

Job turnovers in per cent of total jobs			
Lowest demand on labour		Highest demand on labour market	
Labour office district	%	Labour office district	%
Frankfurt/M.	0.80	Pirna	1.79
Düsseldorf	0.86	Oschatz	1.77
Stuttgart	0.86	Bautzen	1.76
München	0.90	Nordhorn	1.76
Berlin Mitte	0.92	Sangerhausen	1.75
Münster	0.93	Merseburg	1.73
Hannover	0.93	Gelsenkirchen	1.73
Koblenz	0.95	Altenburg	1.72
Hamburg	0.96	Riesa	1.72
Wiesbaden	0.96	Nordhausen	1.69

the results were analysed with regard to the effects in East and West Germany. Here the two scenarios lead to the following results (Table 4.6):

- Within the Ecotax scenario, the increase in employment is in the same order of magnitude between West and East Germany. The increase in East Germany (1.60%) is only slightly higher than in West Germany (1.37%). The labour turnover rate in East Germany amounts to 2.04%. With a labour turnover rate of 1.40% for West Germany, the gross effect is only slightly larger than the net effect. Thus, there are almost no job losses in West Germany, but the increase in jobs is also lower than in East Germany.
- There is a different result within the policy mix scenario: the employment increase in East Germany (0.53%) substantially exceeds the one in West Germany (0.14%). However, it has to be kept in mind that the overall increase in the Ecotax scenario is higher than in the policy mix scenario. The labour turnover rates in East and West Germany differ by about the same margin from the net effect and amount to 1.16% in West Germany and 1.46 in East Germany, respectively.

This difference between the two scenarios can be partially explained by which sectors gain and which lose. In the Ecotax scenario, some service and government sectors benefit from the policy. Taken together, they are distributed equally between West and East Germany (including Berlin).[6] In contrast, construction-related sectors are the ones which stand to benefit the most under the policy mix scenario. However, in relative terms, these sectors are much more important for the East German economy than for the West German one. This results in an above average increase of employment in East Germany.

Table 4.6 Comparison of effects in East and West Germany

	Ecotax scenario		Policy mix scenario	
	Net effect in %	Gross effect in %	Net effect in %	Gross effect in %
East Germany	1.60	2.04	0.53	1.48
West Germany	1.37	1.40	0.14	1.16

Table 4.7 Results for the Herfindahl Index of regional concentration of employment

Herfindahl Index: Reference scenario	13.06
Herfindahl Index: Ecotax	13.22
Herfindahl Index: Policy mix	13.02

4.3.5 Effects on the Regional Concentration of Employment

Employment in Germany is not evenly distributed. There are some districts which accumulate employment, and others with a rather low number of jobs. In order to measure the regional concentration, a Herfindahl Index (HI) of regional concentration has been developed (see Walz et al. 2001). It is normalised in such a way that if the regional concentration in a sector is distributed equally among all regions, it yields a value of 5.22. However, if the employment in a sector is taking place in one region only, the index yields a value of 1,000. The overall Herfindahl Index value for the whole of Germany is calculated by summing up the values for each sector and weighting it by the share of the sector's employment in total employment.

The calculations of the Herfindahl Index for the regional concentration of employment do not vary much between the scenarios. The value of the Herfindahl Index increases slightly from 13.07 in the reference case to 13.22 in the Ecotax scenario. It falls slightly in the policy mix scenario to 13.02. Thus, it can be concluded that only small changes in regional concentration are taking place (Table 4.7).

4.4 Changes in Qualification and Working Conditions

The sectoral changes also imply different effects on the job characteristics and qualification requirements. In order to account for these structural adjustments, the ISIS model was used again. Based on the German microcensus, it contains a submodule which describes the job characteristics and qualification requirements within each of the economic sectors. Thus, the differences between the reference scenario and the two policy scenarios with regard to qualification requirements and

[6] In general the share of government-related jobs is higher in East Germany, the share of service sector-related jobs higher in West Germany.

job characteristics can be calculated using the same data input as for the regional analysis. The following indicators were analysed:

- qualification requirements (master degree, bachelor degree, foreman/technician, apprenticeship, without education/training);
- percentage of part-time jobs and jobs with a limited term job contract, and
- percentage of jobs with an increased need for flexible working hours (weekend/holiday work; evening/night work; shift work).

In the model runs it was assumed that the climate policies do not change the pattern of distribution of the qualification requirements and working conditions within each sector. For the policy mix scenario, this seems immediately plausible because its measures do not influence the cost of labour. Thus, the cost structure of different qualities of labour remains unchanged, too. Within the Ecotax scenario, the cost of labour is reduced, but by a uniform percentage rate. Thus, the costs of highly paid labour (e.g. because of high qualification requirements or night shifts) are reduced by the same percentage as the costs of low wage labour. Therefore it was assumed that neither does the Ecotax scenario change the pattern of qualification requirements and working conditions within each sector.

In order to highlight the effects, the changes in qualification requirements and working conditions in sectors which increase their share in total employment (winners) are contrasted with sectors with a decreased share in total employment (losers). Thus, the actual qualification requirement of the respective scenario is somewhere in-between these two extremes. However, the comparison between losers and winners produces an image of what kind of qualification requirements and working conditions the economy is moving towards within the scenarios, and which pattern of qualification requirement and working conditions is being left behind.

The analysis of the qualification requirements reveals clear differences between the policy mix and the Ecotax scenarios (Fig. 4.5). In the policy mix scenario, there is a shift from high qualification requirements towards medium requirements, whereas the importance of low requirements is hardly affected by the policy. In the Ecotax scenario, in contrast, the qualification requirements are clearly increasing, with a reduction in both lower and medium qualification requirements.

These differences can be traced back to different patterns of changes in sectoral employment in the two scenarios. Among the highest percentages of highly educated labour are found in the government sector and some service sectors. It is in precisely these sectors that the policy mix scenario brings about below average employment changes (losers), the Ecotax scenario, however, results in an above average growth in employment (winners) to some extent.

There are also changes with regard to job characteristics associated with the scenarios (Figs. 4.6 and 4.7). In both scenarios, the number of part-time jobs and the demand for weekend and holiday work are reduced. Both developments reflect the below average share of part-time jobs and weekend and holiday work in the main winning sectors such as construction and certain sectors of the capital goods

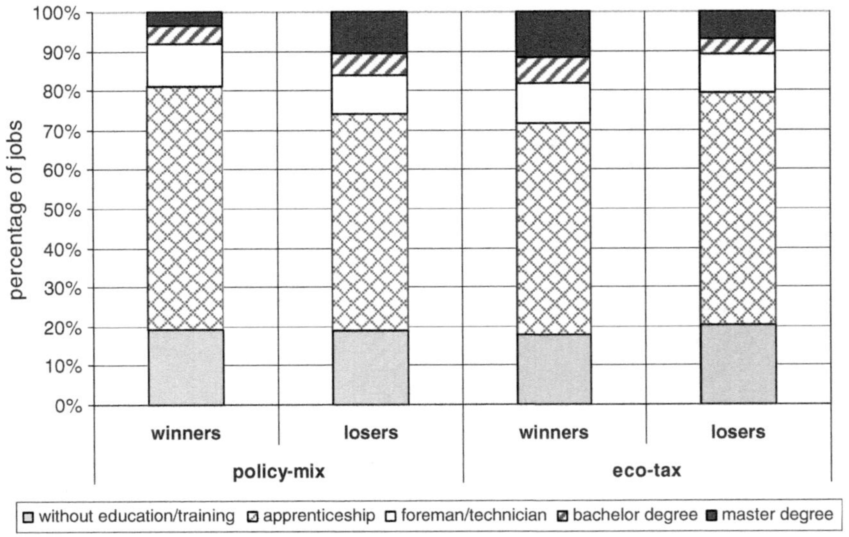

Fig. 4.5 Qualification requirements of winning and losing sectors in the Ecotax and the policy mix scenarios

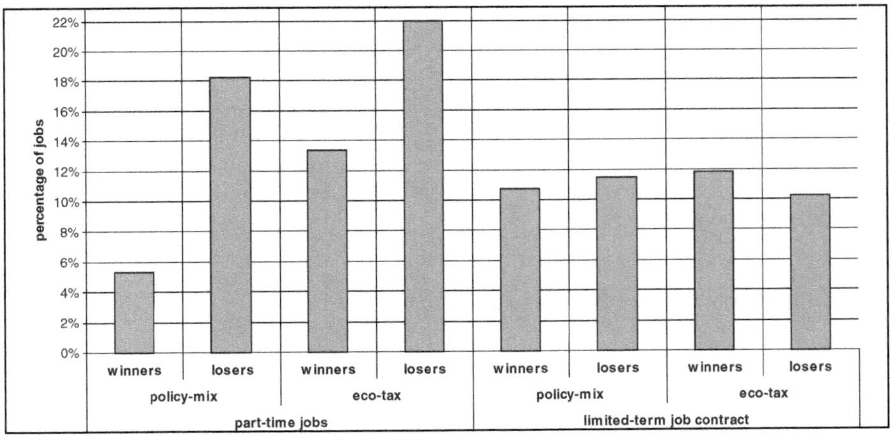

Fig. 4.6 Job characteristics in the Ecotax and policy mix scenario

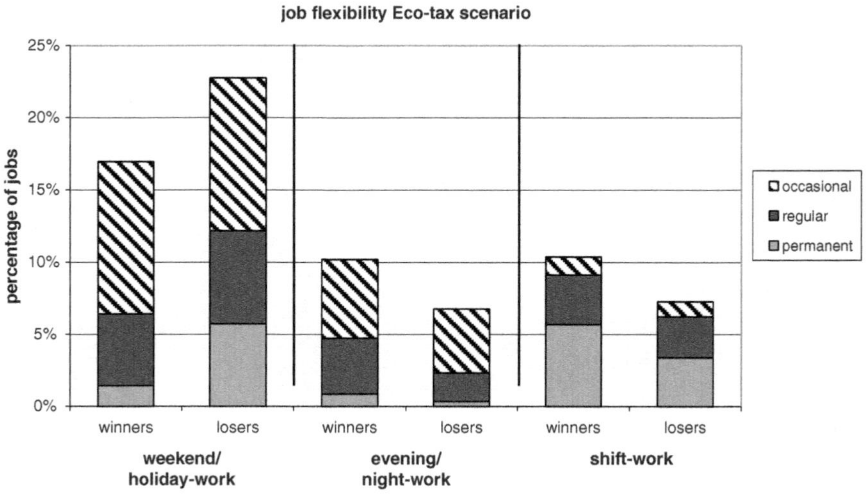

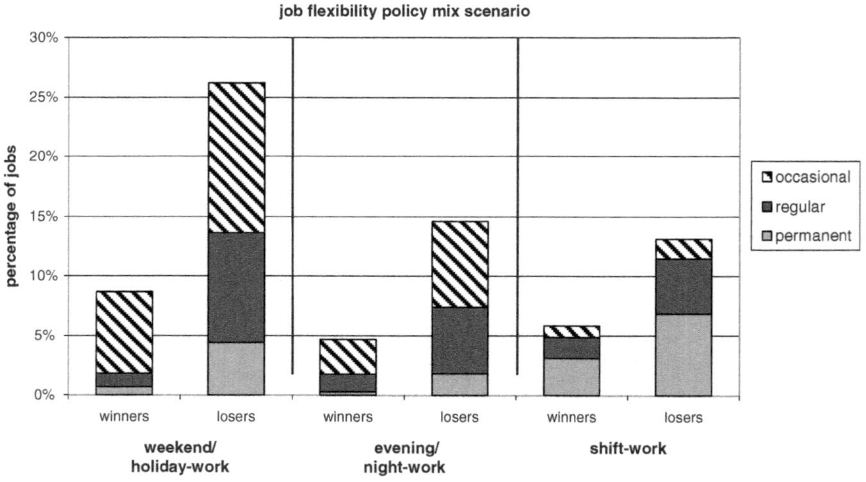

Fig. 4.7 Job flexibility in the Ecotax and policy mix scenarios

industries. The number of limited-term job contracts remains almost constant in the policy mix scenario, but increases slightly in the Ecotax scenario. Furthermore, in the Ecotax scenario, there is a growing demand for night and shift work, but this demand decreases in the policy mix scenario. This difference can be explained by the fact that the most important winning sectors in the policy scenario have below average rates of night and shift work. In the Ecotax scenario, however, some of the winning sectors, e.g. service sectors and car manufacturing, have above average rates of night and shift work. With regard to shift work, the influence of car manufacturing and railways is of overriding significance, leading to an overall increase of shift work in the Ecotax scenario.

Chapter 5
Theoretical Approaches and the State-of-the-Art in Accounting for Technological Change

Chapters 2 and 3 on the macroeconomic effects showed the importance of accounting for technological change in modelling. If technological change towards reducing pollution is induced by environmental policy given environmental targets can be achieved more cost-effectively.[1] Thus, there is an inherent danger of overestimating the economic costs of mitigation if future technology improvements are not taken into account. In particular, it also has to be considered that implementing additional environmental policies can influence the path or the speed of future innovations in itself.

The correlations between environmental protection and innovations have been the subject of more intensive discussion for several years now but still comprise a comparatively new research field when analysing the economic impacts of environmental protection. There are several empirical studies of this subject but these do not cover the entire field and conclusions are ambiguous. Therefore it is necessary, alongside empiricism, to rely on theory to an increased extent in order to derive hypotheses.

This chapter provides an overview of theoretical and empirical approaches to account for technological change. First, the relevant terms are defined in an introductory section. Then various paradigms are presented and discussed with regard to the determinants of innovation: environmental economics, institutional and evolutionary economics, the concept of systems of innovation used in innovation research, regulatory economics and policy analysis. Finally, based on this overview, a short empirical survey on the state of the art is presented.

5.1 Definitions

The importance of the relationship between environmental policies and innovation has grown considerably in the recent past. In contrast to many economic models in which technological change is still regarded as something semi autonomous, that

[1] Goulder and Schneider (1999).

R. Walz and J. Schleich. *The Economics of Climate Change Policies.*
Sustainability and Innovation,

"falls from heaven like manna", there is a widespread consensus in innovation research that innovations can be influenced by numerous modifiable determinants. However, it is essential to first clarify the term innovation. Going back to Schumpeter (1942), the literature traditionally splits technological change into three separate phases of invention, innovation (= first application) and diffusion. This sequential division is being increasingly challenged by more recent innovation research and reference is made instead to evolutionary occurring processes.[2] According to this approach, these three phases do not occur sequentially, but are interlinked in numerous feedback loops. This can be clearly illustrated by the example of users' experiences with alterations from which important clues result for further innovations (this also underlines the significance of the relationships between manufacturers and users).

In this chapter, the term innovation is used in accordance with this more recent interpretation. Thus the main focus concerns the correlations between climate policy and development, first application and the diffusion of new environment-friendly solutions. New environment-friendly solutions are often referred to as environmental innovations. This is a rather fuzzy term. According to Klemmer et al. (1999), environmental innovations comprise of

- technical innovations including organisational ones (new products and production processes, developing new resources and input supplies, changes in corporate structure, business cultures and strategies),
- institutional innovations (reorganisation of social boundary conditions, legal relations and organising principles), and
- social innovations (changes in relevant norms, behaviour and lifestyles),

as far as they – independent of their economic benefit – contribute to improving the quality of the environment. Here, a broad definition of environment is assumed which includes protecting resources (increasing resource efficiency, substituting exhaustible resources by renewable ones), reducing pollution and repairing already incurred environmental damage as well as cross-cutting aspects such as, e.g. the monitoring and diagnosis of environmental pollution. In keeping with this definition, environmental innovations comprise more than just technologies, since they also include organisational, institutional and social innovations.

In innovation research, it is customary to distinguish between incremental and radical (environmental) innovations. Incremental innovations use resources, energy and land more efficiently and alter the possible application areas of already existing systems. These innovations take place within existing paradigms and constitute the biggest share in practice. They are usually less the result of research and development activities (R&D) and more due to the consistent implementation of specific experiences made in the learning processes of engineers and users. Radical innovations, in contrast, compel the replacement of a large share of the existing knowledge, abilities, products, processes and production systems, in short of the capital

[2] See e.g. Grupp (1998), Kemp (1997), Albrecht (2002), Montalvo (2002).

stock. Sweeping innovations can be regarded as a transition type which alters individual products and process stages while retaining existing capital goods to the greatest possible extent.

Alongside the classification into incremental or radical innovations, environment-friendly solutions can also be differentiated by innovation complexity, since the individual environmental innovations require varying degrees of corporate adjustment. For innovations in the field of industrial energy supply, this concerns measures which can be assigned to production-integrated environmental protection. They only affect usually the direct environment of the user of the innovation. However, if changes in the main system are involved, the innovation complexity increases since the complete production process is then affected.

5.2 Neoclassical Innovation and Environmental Economics

There are different paradigms and frameworks to deal with the relationship between environmental policy instruments and innovation effects. One of the most outspoken approaches is environmental economics. Within economics, technological change is broken down into three distinctive phases: invention, innovation, and diffusion. One of the main problems discussed within environmental economics is the choice of the "best" instrument. Among the criteria used, the effects on innovation play an important role and are usually the main theme of the "dynamic efficiency" criteria.

The bulk of the literature concerns the incentives for companies to develop and use new technologies. The main argument derives from the theory of induced innovation which is based upon Hicks (1932): changing relative prices induce innovations which result in the substitution of the production factor becoming relatively more expensive. Thus, an increased price for energy use not only gives a clear signal for the diffusion of efficient technologies, leading to an upward shift of the traditional diffusion curve; usually, an energy tax also increases the average price of energy. Thus, compared to a reference case without an energy tax, there is an additional incentive for innovation even after the most efficient technology currently available has been installed. Thus, it is assumed that changing costs for energy use also lead to incentives for future inventions and innovations.

Newell et al. (1999) generalised this model by including regulating standards. In this way, non-price-related restrictions can be analysed within the scope of the induced-innovation hypothesis if their impacts can be interpreted as changes in shadow prices or implicit prices.

Arrow (1962) was the first to include learning effects in the analysis of economic growth. Learning effects in production process, for example, imply that the specific labour input per unit of capital decreases with the age of capital vintages. Thus, investments not only improve the productivity of the present capital stock but, since they generate new knowledge, they also increase future productivity. In empirical economic models, this type of technological change is often captured through specific costs, which (negatively) depend on accumulated capacity. Here, accumulated capacity represents knowledge which was generated during production (learning-by-doing) and application (learning-by-using).[3]

In models based on the so-called "New Growth Theory", which has been developed by Lucas (1988), Romer (1990), Barro (1990) and Grossman and Helpman (1990), endogenous technological change emerges as the result of public and, in particular, private R&D investments. Investments in R&D not only benefit the investor firm, but increase the productivity or the product quality of all other firms as well. Eventually, long-term economic growth is only feasible because of these so-called spillover effects.

A crucial assumption in these predominantly neo-classical models of innovation is that decision makers act perfectly rationally (homo economicus): investment decisions are based on a profound analysis of the benefits and costs in order to maximise profits, while cost and benefit figures are perfectly known in terms of their expected values. However, these assumptions appear questionable since innovations are, by nature, the result of a series of unpredictable events.

To conclude, in modern approaches to technological change, innovation is considered to be the result of economic activities which require the use of scarce resources. This type of endogenous/induced technological change may be generated through private or public R&D, education, learning effects, spillovers, or price changes. The new growth theory's diverse explanations for technological change have important ramifications for policy intervention. While, by nature, policy cannot affect autonomous technological change, properly designed policy measures may be well suited to spur endogenous technological change (via, for example, measures to promote R&D and spillovers).

From the hypothesis of induced innovation, conclusions can be derived for the effects of various policy instruments on innovation, in particular on the diffusion of new technologies.[4] It is decisive for the innovation effect that the instruments emanate a continuous financial incentive. Here, neoclassical environmental economics rate economic instruments as the best: "The most comprehensive effects on the advance of environmental technology result from levies and tradable emission certificates, since the cost burden on the remaining emissions produces a permanent incentive to search for advanced possibilities of emission reduction. This triggers not only cost-reducing but also emission-reducing disposal technologies and integrated environmentally-friendly production processes."[5] In contrast to this, the use of command-and-control policies is predominantly judged as not very innovation-friendly. True, there is an incentive to comply with the given thresholds more cost-effectively using cost-reducing innovations, but there are no incentives to do more

[3] For overviews on the modelling of technological change in environment-economic models see the surveys by Löschel (2002) and Carraro and Galeotti (2002).

[4] See Fischer and Newell (2004) for a recent theoretical and stylized empirical assessment of various climate policy options. The few empirical studies on the benefits of market-based policy instruments include Kerr and Newell (2001) and Newell and Rogers (2003).

[5] Michaelis (1996, p. 48).

than is necessary for environmental protection, since no costs arise for the remaining (permitted) pollution. An incentive for advanced innovations is conceivable for the manufacturers of environmental technology if the requirements are expected to become more stringent. This is the case, e.g. if the thresholds set are oriented on a best available technology which insures at the same time a quasi state-guaranteed minimum demand for the new environmental technologies. This conflicts with the fact that the users affected have an incentive not to make known existing possibilities for further reduction of the pollution. This conundrum has become known in the literature as "the chief engineers' code of silence".[6]

To sum up, environmental economics moves within a rather linear model of sequential innovation stages: inventions lead to new technical development, which then diffuses through the market. There is a tendency to analyse the effects on the different stages of innovation separately. Assuming perfect economic rationality, innovation decisions are based on microeconomic optimising behaviour. As a result, environmental economics sees more positive effects from market-based instruments on the development of new technologies.

5.3 Evolutionary and Institutional Economics

Evolutionary economics, which explicitly allows for the future openness of innovative processes, offers promising approaches to explain technological change.[7] It regards innovative processes from a novel perspective and uses the natural science analogy of open and closed systems as well as biological evolution to do so. Equilibrium states may evolve in closed systems without links to the outside world. In open systems, interaction with and reactions to the environment occur (for example by exchanging information). Descriptions of state – also for the so-called stationary states – are only valid temporarily. Two mechanisms are seen as being central to the emergence of innovations: the generation of variety and selection. A larger variety per se is conducive to (environmental) innovations.

As far as certain developments have created favourable conditions for economic and technical change, an irreversible transition to new states then occurs through the use of temporary windows of opportunity. The fact that the developments being triggered are not predictable justifies the assumption of the "future openness of innovative processes". Grupp (1998) cites as examples inventions, discoveries, new organisations, the movement of human capital within or between sectors, changes in values and new conditions for competition. This paper focuses particularly on the favourable conditions mentioned (such as innovation-friendly frame conditions), so

[6] See Michaelis (1996). However, the case study by Wallace (1995) of the German SO_2 reduction programme reveals how such a code might be broken.

[7] There are various schools among the paradigm of evolutionary economics. See, e.g. Nelson and Winter (1982), Dosi (1982), Dosi et al. (1988), Erdmann (1993), Nelson (1995), Grupp (1998), Weber (1999), Blazejczak et al. (1999), Witt (2003).

that innovation and diffusion can positively influence each other in a sort of feedback process (learning process).

With regard to behaviour, the strict rationality of the "homo economicus" is softened even if selection processes can result in empirically observed behaviour acting "as if" it were rational. Nelson (2002) stresses the point that behavioural routines which have evolved over a longer period of time play an important role, and take the place of the permanent optimisation due to smallest modifications in the frame conditions which dominates neoclassical theory. This behavioural assumption is implicitly linked with a restriction of the induced innovation hypothesis based on relative price changes of neoclassical theory and the resulting instrument preferences. If innovation behaviour is determined by behavioural routines, not only changes due to altered relative prices are decisive, but also changes in the behavioural routines themselves. Thus, voluntary agreements, which are often criticised for their lack of stringency (Jochem and Eichhammer, 1999), may nevertheless lead to a change in behavioural routines.

New institutional economics emphasises the fundamental significance of institutions for all aspects of economic behaviour.[8] Richter (1994, p. 2) defines the term institution as "a system of norms, including their guarantee instruments, geared to a specific target with the purpose of steering individual behaviour in a specific direction [...]. Institutions can be formal in the sense of objective and subjective rights and informal". In an extreme case, they can emerge "spontaneously", i.e. organise themselves, or be completely structured by an authority. The assumption of bounded rationality, which was originally developed by Simon (1957), is closely linked to this approach, as is the assumption of opportunistic behaviour and allowing for information costs. These considerations highlight the importance of institutional arrangements in such analyses.

The following aspects are of particular significance with regard to environmental innovations:[9]

- Groups of actors, often with opposing interests, are involved in the process of developing environmental policy plans (state, industry, associations, NGOs). It can be assumed that each group is characterised by individual types of behaviour and the attitude of its members. Of interest is how the necessary collective action is achieved and how compromises prove to be good solutions in practice.
- Transaction costs play a considerable role and have to be accounted for when designing environmental policy instruments. These include resources necessary for the creation, maintenance, support and equipment of institutions and organisations. In addition, search and information costs occur, negotiation and decision costs and monitoring and implementation costs as soon as players become active in markets. High transaction costs can act as drivers for and barriers to technical and organisational or institutional innovations.

[8] See Coase (1991), Williamson (1975, 1985), Williamson and Winter (1991), Eggertsson (1990), Richter and Furubotn (1999).

[9] See Richter (1994), Klemmer et al. (1999), Ostertag (2003).

- Property rights: institutional economics distinguishes between absolute rights (rights to things, immaterial rights), relative rights (between two parties) and individual rights to freedom. Safeguarding, assigning and using property rights incur costs.
- Incentive systems: finally it can be assumed that environmental innovations are realised and promoted only via particular incentive systems. These have to incorporate a target system (e.g. improved environmental quality, sustainable development) and fixed rules which oblige individuals to behave in concrete ways and which incorporate their individual objectives. Controls and possible sanctions have to be defined and conveyed to the actors via an information system.

5.4 Systems of Innovation Approach

The approaches of evolutionary and institutional economics have also influenced the more empirically-oriented innovation research. In the 1990s, the heuristic approach of systems of innovation gained wide acceptance.[10] In addition to the demand and technology factors, this approach underlines the manifold aspects of the intra-firm determinants of innovation, the characteristics of innovation as an interactive approach, the role of institutions in shaping activities, the importance of the home (lead) market as a base for competitiveness on the international markets, and the regulatory framework. The key notion of the systems of innovation approach is that these factors influence each other, highlighting the importance of feedback mechanisms (see Fig. 5.1). This results in an expansion of the influencing factors. As well as players estimating the profitability of innovations, increased significance is being attached to soft context factors such as, e.g. communication patterns between the participants, but also the regulatory pattern between policy and those governed.[11]

Within the concept of innovation systems, environmental policy instruments have scarcely been used as a determinant for innovation.[12] However, it is easy to see that such instruments could be incorporated into this framework as an additional factor for innovation, e.g. under the heading of the role of the home market (also prominent as a first mover advantage in the work of Porter 1990) and as an important feature of the regulatory framework. Within the innovation systems approach, the analysis is always very context specific and the effects of the various factors depend on the systems' conditions. Thus, there is no clear hypothesis about the specific effects of the various environmental policy instruments.

[10] For excellent reviews, see Edquist and McKelvey (2000), Carlsson et al. (2002), Lundvall et al. (2002), Edquist (2005).

[11] See SRU (2002), Leone and Hemmelskamp (2000), Kemp et al. (2000), Montalvo (2002).

[12] There are some studies dealing with the introduction of renewable energy, such as Bergek and Jacobsson (2003), Walz and Kotz (2003).

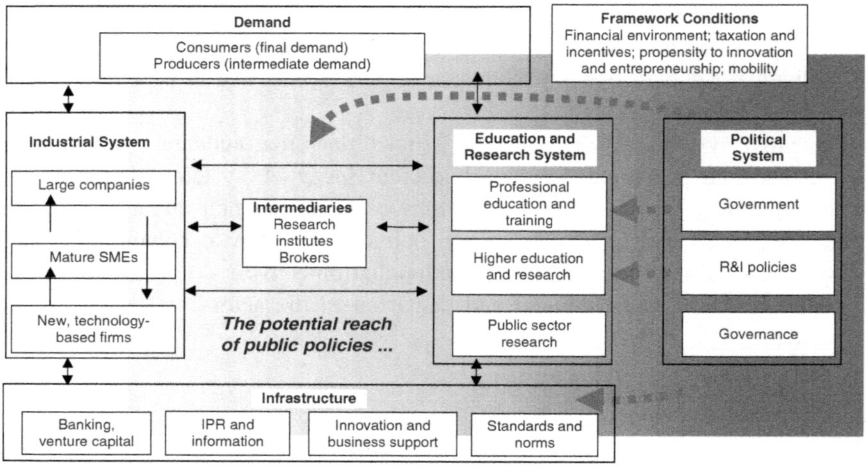

Fig. 5.1 Heuristic diagram of an innovation system (Source: Kuhlmann and Arnold 2001, p. 6)

The framework of systems of innovation has been applied traditionally to national innovation systems. More recently, however, it has been also applied to analyse technological or sectoral systems.[13] These approaches share the starting point that innovations can be best explained by characterising the components of an innovation system, such as actors, networks and institutions, and their interaction with each other. Furthermore, it has been suggested that a technological innovation system can be best analysed by looking at how the different functions an innovation system has to meet are fulfilled.[14] There is no final list yet, however the following functions can be distinguished:

- creation of new knowledge,
- creation of positive external economies through exchange of information and knowledge,
- demand articulation, including guidance with respect to technological and market choice,
- recognition of a growth potential, which is closely connected to the legitimacy of a new technology,
- facilitation of market formation,
- supply of resources, which is especially important for new technologies which are associated with a higher risk of failure, and
- arenas for coalition building and organisation of interests, and alignment of competing interests by offering a platform for negotiation.

[13] See e.g. Carlsson and Stankiewicz (1995), Carlsson et al. (2002), Malerba (2002, 2005).
[14] Johnson (1998), Jacobsson and Johnson (2000), Bergek and Jacobsson (2003), Smits and Kuhlmann (2004).

The development of an industry goes through different phases.[15] In general, a first phase of experimentation with frequent entries and exits, many different technological alternatives, and a small market, can be distinguished from a second phase, which is characterised by market growth and consolidation of the suppliers (fewer new entrants, concentration of suppliers). Bergek and Johnson (2003) suggest that the importance of the different functions of an innovation system varies. In the first phase of experimentation, creation of new knowledge and guiding the search process are very important. This requires the creation of both legitimacy of and variety between the technological approaches, and the entry of new actors and the creation of networks to ensure positive external economies. In the phase of market growth, the formation of a mass market becomes the key prerequisite, emphasising also the function of supply of resources.

To sum up, the systems of innovation approach has gained wide acceptance among innovation researchers in the past. Lately, this concept has been further developed by looking at detailed functions of an innovation system, and by scaling its aggregation level down to the sectoral or technological level. Both developments open up the perspective to analyse the innovation effects of specific energy and environmental regulations integrated into the wider framework of a system of innovation approach.

5.5 Regulatory Economics

The innovation of technologies in the energy field depends very much on the strategies of the major players in the market. With regard to climate change, the electric and gas utilities clearly belong to the key players. However, in addition to environmental regulatory challenges, these actors are also subject to one form or another of specific economic regulation of the sector.

The traditional case for regulation of public utilities was the existence of a natural monopoly, resulting in a rate-of-return regulation or in some form of cost-based pricing. In relation to innovation, theoretical work has shown that rate-based regulatory schemes can result in a biased technical change towards capital intensive production.[16] During the last two decades, regulatory economics started to emphasise the need to consider the incentive scheme within regulation. Based on the progress of the economics of information, the information asymmetries between regulators and regulated companies was addressed by Laffont and Tirole (1986) as a principal agent problem, leading to the implementation of new regulatory schemes such as "price cap" or "incentive" regulation.

The most important change in regulatory economics has been the call for deregulation. Theoretical insights of the theory of contestable markets, e.g. from Panzar

[15] Nelson (1994), Utterback (1994).

[16] This impact has become known as Averch-Johnson Effect. See Averch and Johnson (1962), Zajac (1970).

and Willig (1977) or Baumol (1982), led to the conclusion that only monopolistic bottlenecks characterised by both sunk cost and natural monopoly cost functions should be regulated. Clearly, infrastructure systems based on physical networks such as electricity/gas, water supply and sewage treatment, or railways include such a monopolistic bottleneck. However, among the four vertical stages of electricity supply, only transmission and distribution fulfill the conditions of a monopolistic bottleneck, not however the other two stages, generation of electricity and retail to the end-users, which are (potentially) competitive. The potentially competitive stages, in general, require access to the monopolistic bottlenecks. This also holds for power produced by independent power producers or for new players such as operators of wind turbines.

The situation is complicated further, if there are vertically integrated utilities which are active in both, the monopolistic bottlenecks and the potentially competitive stages. In this case, regulation has to deal with the problem that the market power within the monopolistic bottlenecks can be carried on to the potentially competitive stages either by excessive charges for access to the monopolistic bottlenecks, or by hindering or even foreclosing the downstream market to competitors.[17] As a result, there is no level playing field between incumbent utilities and newcomers such as independent power producers or wind turbine operators. To sum up the theoretical arguments, access to the grid plays a very important role for the development of electricity generation technologies which challenge the mainstream of the incumbent utilities technology choice. Furthermore, the design of regulation in a liberalised electricity market has considerable effect on the ability of incumbent utilities to pursue market strategies which deter the entrance of newcomers.

5.6 Policy Analysis

Another view of the effects of environmental policy has been developed by political scientists within the so-called "policy analysis paradigm".[18] This paradigm structures the policy process in five stages: perception of problem, agenda setting, implementation, monitoring, and evaluation. In contrast to environmental economics, this paradigm sees the determinants of the success of an environmental policy not so much within the stage of implementation, but rather within the perception of environmental problems and the agenda setting. The importance of the choice of instrument is downplayed.[19] Instead, other factors are viewed as being of much

[17] Knieps (2001).

[18] See Héritier (1993), Howlett and Ramesh (1995), Jänicke et al. (1999).

[19] Jänicke and Weidner (1995), Jänicke et al. (1999), Jänicke (1996, pp. 11–12; 1999, pp. 107–108) argues that the idea that specific instruments have specific effects is misleading in most cases. Very often not the choice of the instrument is important, but that an action is taken at all, regardless of which. The communication patterns are decisive. If environmental target finding is consensual, reaching the target is highly probable – independent of the instruments used to do so.

greater significance, such as the power and strategic ability of the various players, the nature of the problem, political environment, and policy style.

Policy style, in particular, can be influenced by policy makers. Richardson (1982) and Jänicke (1996) point out that empirical analysis has shown the importance of interaction between state and private players. In a complex world, policies should be part of a learning process. A precondition for such cooperation is a policy style which enables a dialogue between those involved. Furthermore, firms need a certain degree of reliability if they are to engage in innovative activities. With regard to environmental innovations, this requires that the political priorities of the environmental problems are known in advance. Thus, the existence of a long-term policy plan clearly stating the environmental medium- and long-term goals is viewed as being one key factor for an innovation-friendly environmental policy. To sum up, policy analysis highlights the importance of the soft context factors and downplays the importance of the choice of instrument. To this extent, the approach clearly takes a counter position to environmental economics.

5.7 Review of Empirical Results

In their review of the hypothesis of induced innovations, Thirtle and Rutan (1987) concluded that it can be inferred from existing statistical surveys that changing the relative factor prices does indeed have an impact on the pace of innovation. However, the studies assessed did not concern the environmental domain. In the environmental domain, namely, there is a major problem with transforming the changes, which are strongly characterised by command-and-control type measures, into statistically measurable variables. Accordingly, there are far fewer econometric-statistical analyses available, which, in addition, frequently have to use proxies as explanatory variables.

One important hypothesis which has been analysed is the effect of the stringency of the environmental policy on innovations. Various studies used environmental spending and patents as proxies for these variables. Landjouw and Mody (1996) and Grupp (1999) conclude positive correlations between environmental spending and patent activity in the related technology fields. However, this correlation is not confirmed by Jaffe and Palmer (1997). In a more recent study, DeVries and Withagen (2005) find a strong positive relationship of high emission levels as a latent variable for strict environmental policy with environmental innovation.

The effect of environmental management systems is also a subject of discussion in the literature. In their review of studies from the nineties, Dyllick and Hamschmidt (1999) summarise that environmental management systems are strongly geared towards short-term process controls and less towards innovations. In contrast, in a study by Rennings et al. (2003) environmental management systems under EMAS (Eco-Management and Audit Scheme) are assigned a positive effect on the realisation of environmental innovations. Wagner (2007) found that environmental management systems do trigger process innovations, but not product innovation.

There have been various studies recently using survey results for analysing the influence of various determinants of environmental innovations.[20] Rennings et al. (2006) and Rehfeld et al. (2007) show that organisational aspects are of importance. Arimura et al. (2007) and Frondel et al. (2007) support the hypotheses that the policy stringency are important factors. Horbach (2007) stresses the point that different factors – environmental policy, management tools and general innovation capability of the firms – all contribute to environmental innovation.

The most intensive examination so far has been made of the field of energy, especially as a statistically sound, explanatory variable is available here in the form of energy prices. The more recent studies from Newell et al. (1999), Grupp (1999), Schleich (2001), Popp (2002), and Lutz et al. (2005) suggest that increases in the relative energy prices trigger energy-saving innovations. But the statistical significance of this correlation varies as does the magnitude of the influence of the energy prices. In addition, there is a large body of literature breaking down the changes in aggregate energy intensity – or more recently – carbon intensity, into changes in the structure of the economy and changes at the level of sub-sectors.[21] However, these studies tend to be purely descriptive, and no attempt is made to explore the determinants of the observed changes, such as price changes.[22] Thus, there is a need for additional research, which combines both approaches, in order to come up with a more sound analysis on the influence of energy prices.

Overall, it can be concluded from this work that relative changes in environmentally-relevant costs actually do influence environmental technology development and the diffusion of energy-efficient technologies, but because a series of other case-specific determinants exists for each individual subject of investigation, it is not possible to put forward a generalised quantitative relationship for inducing environmental technology progress. In particular, the impact of energy costs on innovation varies between sectors because there are differences in the respective technologies, market structures, organisational culture, or regulatory environment involved.

Alongside econometric analyses there are also case studies examining the correlations between environmental protection measures and environmental innovations. The work of Porter and van der Linde (1995) is often cited in the literature. They indicate considerable innovation effects and even the existence of an extensive unexploited efficiency potential, whose realisation could result in a so-called win-win situation, in which environmental protection actually precipitates a reduction of the microeconomic cost burden.

The implementation of technologies and practices which reduce energy consumption at the level of private and public organisations or individual households

---

[20] See Horbach (2007) for an overview.

[21] See e.g. Unander et al. (1999), Diekmann et al. (1999), Schipper et al. (2001), Liaskas et al. (2000).

[22] Recent exceptions include Miketa (2001) and Welsch (2001).

is often hindered by obstacles. Many of these measures are considered cost-effective from the company's or individual's perspective under prevailing economic conditions.[23] There is a substantial body of literature that analyses the nature of these barriers which are, in essence, due to market failures, caused by high information costs and other transaction costs, hidden costs, financial or technological risks, capital market restrictions, split-incentives (landlord/tenant dilemma), as well as organisational and behavioural constraints (Brown, 2001; Eyre, 1997; Howarth and Andersson, 1993; Jaffe and Stavins, 1994a, b; Ostertag, 2002; Sorrell et al., 2004; Stern, 1986). Empirical analyses of the relevance of the various types of barriers and of the determinants often rely on theory-based case studies (DeCanio, 1994; De Almeida, 1998; InterSEE, 1998; Ramesohl, 1998; Schleich et al., 2001a; Sorrell et al., 2004; Sorrell, 2003). Such case studies are well suited for gaining insights into complex decision-making processes and structures within organisations. Yet, the empirical basis for a generalisation of both their findings and their policy recommendations is rather weak.

Due to a lack of data, there are only a few studies on the relevance of barriers to the diffusion of energy efficiency that are based on ample empirical evidence. While performing univariate analyses, Gruber and Brand (1991), and Jochem and Gruber (1990) focus on decision-making in companies of the commercial sector with regard to energy efficiency measures. For other sectors, few multivariate econometric analyses have been carried out. Brechling and Smith (1994) examine the take-up of wall insulation, loft insulation, and double glazing in the UK household sector. For Irish households, Scott (1997) carries out a similar study looking at attic insulation, hot water cylinder insulation and low energy light bulbs. For the industry sector, DeCanio (1998) investigates companies' investment behaviour based on data from the United States Environmental Protection Agency's Green Lights programme. Finally, DeGroot et al. (2001) analyse to what extent barriers to the implementation of energy-saving technologies in Dutch companies vary across sectors and firms' characteristics, running separate regressions for each potential barrier. Finally, Schleich and Gruber (2008) econometrically assessed the relevance of several types of barriers to energy efficiency for the German commercial and services sector. None of these papers, however, econometrically explore the impact of policies on barriers to energy efficiency.

Positive impacts of environmental command and control policies on innovations were shown for several European countries in the case studies of Wallace (1995). Particular attention should be drawn to the research programme "Innovative effects of environmental policy instruments (FIU)" of the German BMBF.[24] Different environmental policy measures were examined here. It was shown that command-and-control type measures can also have a positive innovation effect, but that a multitude of system requirements have to be considered which makes it difficult to generalise the results. Most importantly, the results – which are emphasised among others by

[23] See Sect. 2.2.2 for a discussion of the so-called no-regret options.

[24] For a summary of the results of this research programme, see Klemmer (1999).

the evolutionary and institutional approaches and environmental policy analysis – indicate that the impact of the system requirements and soft context factors is not negligible. On the other hand, the case studies also show that price expectations are particularly significant in the context of the frame conditions which is, in turn, consistent with the hypotheses of neoclassical environmental economics. This implies that the theories outlined in the preceding paragraphs should not be interpreted as opposing but rather as complementing each other. Results from the follow-up programme "Frame conditions for innovative economies" (RIW) of the German BMBF seem to confirm this conclusion.[25]

To sum up, various case studies indicate that soft context factors and the reduction of obstacles are key factors for innovation especially in the non-energy intensive sectors. However, there clearly is a need to move from the case study based empirical research towards a wider empirical basis. This holds especially for the analysis of various policy measures and the influence they have on innovations.[26]

There has not been much empirical work on the influence of different regulatory designs on technological innovation in the energy sector. However, the work of Walz (1995b, 2002) suggests that even minor details in the regulatory design may trigger important effects on innovation. Even details such as the provisions for allowing construction work in progress or overcapacity to be considered in the rate base can lead to substantial effects on electricity generation technology, by either hindering or favoring the development of capital intensive technologies, or by allowing the buildup of overcapacity which can be used as a strategy to deter newcomers from entering the market.

Recently, there have been various case studies in the field of renewable energies. Most of them focus on the debate about the incentive structure of different policy instruments.[27] These studies conclude that policy measures were a considerable driver for innovation by stimulation R&D through subsidisation programmes and by regulating the feed-in prices for electricity produced by renewable energy sources. Mandatory fixed feed-in payments seem to produce much greater effects than quota or bidding systems mainly because the feed-in tariffs which, in the case of Germany, are fixed for 20 years provided a more stable planning horizon. A few studies also draw either on the policy analysis approach or on evolutionary economics.[28] They conclude that there is also a succession of other system requirements which are significant for success, such as the communication between the players and the existence of long-term policy goals which foster the development of a manufacturing industry for renewable energy sources and thus justify initial public innovation efforts. However, these studies, by and large, do not analyse the effects on innovation within an integrated systems of innovation view. Within the

[25] Horbach (2005).
[26] See Chap. 7.
[27] See e.g. Haas et al. (2004) and Ragwitz et al. (2005a) with overviews for Europe.
[28] Reiche and Bechberger (2004), Lauber and Mez (2004) and Markard et al. (2004), respectively.

concept of innovation systems, on the other hand, environmental policy instruments have scarcely been used as a determinant for innovation. And even the few case studies from this research tradition, which deal with energy issues,[29] do not go into detail with regard to the electricity specific forms of regulation (e.g. liberalisation) or the specific role of the various energy policy instruments. Thus, there clearly is a need – and an opportunity – to merge these two research traditions, by integrating research on the specific effects of regulation design into a wider system of innovation approach.[30]

[29] See Jacobsson and Johnson (2000), Bergek and Jacobsson (2003), Agterbosch et al. (2004), Foxon et al. (2005).
[30] See Chap. 8.

Chapter 6
Technological Change and Energy Consumption in Energy Intensive Industry

6.1 Technological Change and Energy Consumption in the German Manufacturing Sector

In this section, the relationship between technological change and energy use in the German manufacturing sector is explored empirically. In particular, the impact of changes in energy prices is assessed. First, the energy consumption of the German manufacturing sector is presented. Then econometric methods are applied to explore to what extent the observed decline in fuel intensity in the German manufacturing sector can be attributed to changes in the fuel prices over this period.[1] Finally, using data for West Germany between 1970 and 1994, the total fuel intensity of the manufacturing sector is decomposed into structural change and efficiency effects.[2]

6.1.1 Energy Consumption of the German Manufacturing Sector

Between 1991 and 2002, final energy consumption in German industry (manufacturing and other mining sectors) decreased by about 13% from almost 2,700 to below 2,350 PJ. At the same time gross value added grew by about 11% in real terms.[3] Thus, over that period of time, energy intensity in the German manufacturing sector, as measured by the ratio of final energy consumption to gross value added, declined by about 22%. Although a significant part of this decline was due to the restructuring processes and efficiency gains in the new federal states following

[1] Equivalent econometric analyses for electricity consumption failed primarily because there was only little variation in electricity prices over this period.

[2] Due to a lack of data, analyses for longer time periods were not feasible.

[3] Arbeitsgemeinschaft Energiebilanzen (2003), Statistisches Bundesamt (2003).

German unification in 1990,[4] the observed decline is part of a long-term trend: final energy consumption in the German industry sector today is roughly one third below its 1970 value. Likewise, industry's share in final energy consumption in Germany has declined steadily from more than 30% in the early 1990s to about 25% today. As for the composition of energy consumption, only fuel consumption declined, while electricity consumption increased significantly. Compared to 1970, German industry has increased electricity consumption by more than 30% (more than 50% in West Germany), while fuel consumption has dropped by more than 45% (more than 30% in West Germany). As for energy carriers, the share of gas increased in the 1990s from about 20% to over 50%, while the share of mineral oil and hard coal declined.[5] This fuel switch has also resulted in a significant decline of CO_2 emissions and thus contributed towards achieving national and international greenhouse gas emission targets.

From a descriptive perspective, changes in industrial energy consumption are not only determined by the *activity* level, that is, by production in the industrial sector, but also by structural composition. Economic as well as technological factors may result in *inter-sectoral* structural change from energy-intensive sub-sectors towards less energy-intensive sub-sectors. Ceteris paribus, such a shift will result in lower energy consumption for the entire sector. The significance of inter-sectoral variation in the energy consumption of German industry is illustrated in Table 6.1.

In Germany, only four energy-intensive sub-sectors, i.e. minerals, iron and steel, basic chemicals, and pulp and paper, account for almost 2/3 of total final

Table 6.1 Sectoral final energy consumption in the German manufacturing sector in 1999

Sector	Final energy consumption		Fuel consumption		Electricity consumption	
	PJ	Share in %	PJ	Share in %	PJ	Share in %
Minerals	216.3	9.1	187.2	11.3	29.1	4.0
Iron and steel	552.5	23.2	487.0	29.3	65.5	9.1
Non-ferrous metals	132.3	5.6	61.1	3.7	71.2	9.9
Basic chemicals	364.3	25.3	222.1	13.4	142.3	19.7
(other chemicals)	(110.6)	(4.6)	(78.1)	(4.7)	(32.6)	(4.5)
Pulp and paper	172.9	7.3	111.2	6.7	61.8	8.5
Glass and ceramics	98.0	4.1	79.0	4.8	18.9	2.6
Food	185.4	7.8	137.4	8.3	48.0	6.6
Others	551.4	23.1	297.7	17.9	253.7	35.1
Total[a]	2383.9	100.0	1660.9	100.0	723.0	100.0

Source: Arbeitsgemeinschaft Energiebilanzen (1999)

[a] Manufacturing sectors including other mining, without oil processing industry

[4] In the first 2 years after the fall of the Berlin Wall, gross value added in the manufacturing sector decreased by over 60% (Ziesing et al., 1997). Consequently, about half the greenhouse gas emissions reductions observed in Germany in the 1990s can be attributed to this "wall-fall" effect (Schleich et al., 2001b).

[5] Arbeitsgemeinschaft Energiebilanzen (1999).

energy consumption in the industrial sector. Likewise, *intra-sectoral* structural change, that is output or product-mix changes within the same sub-sector or branch, will generally result in changes in energy consumption. In addition to structural change, energy consumption is affected by *technological progress*, which includes improved energy efficiency for existing processes, changes in the mix of energy carriers (substitution of waste for fossil fuels), recycling, or process substitution.

There is a large body of literature breaking down the changes in aggregate energy intensity – or more recently – carbon intensity, into changes in the structure of the economy and changes at the level of sub-sectors.[6] However, these studies tend to be purely descriptive, and no attempt is made to explore the determinants of the observed changes. Recent exceptions include Miketa (2001) and Welsch (2001).

From a policy perspective, the question arises whether some of the observed decline in energy intensity can be attributed to energy prices. Of particular interest is whether technical progress is autonomous or whether it is the result of resource spending. In the latter case, technological progress is said to be endogenous. In contrast to autonomous technical change, endogenous technical change may be affected by policy interventions. According to the theory of *induced innovation* developed by Hicks (1932), changes in relative factor prices will result in innovations requiring less of the more expensive factor. Thus policies such as energy or carbon taxes, which increase the price of energy or carbon, not only result in a different factor mix for the existing production set, but also lead to the invention of new, more energy-efficient technologies. However, there is only limited empirical evidence on the link between environmental policies and technological change. In addition, most empirical economic models do not allow for endogenous technological change.[7]

In general, the observed drop in industry's energy intensity over time may be the result of (1) inter-sectoral structural change; (2) intra-sectoral structural change; (3) factor substitution and improved energy efficiency within the existing production possibilities, and (4) the adoption and diffusion of new, more energy-efficient technologies. The latter is likely to be particularly relevant in many energy-intensive sectors where factor substitution is limited. In these cases, technologies tend to be of the putty-clay type, that is, companies may chose from among different technologies (technological paradigms) when making investment decisions, but thereafter, according to Dosi (1982, 1988), the input structure is basically fixed. Gilchrist and Williams (2000) estimate the share of putty-clay technologies in total output as ranging between 50% and 70%. They consider this share to be even higher for energy-intensive sectors.

---

[6] See e.g. Unander et al. (1999), Diekmann et al. (1999), Schipper et al. (2001), Liaskas et al. (2000).

[7] Exceptions include Goulder and Schneider (1999), Goulder and Matthai (2000), Buonanno et al. (2003) or Lutz et al. (2005). For comprehensive surveys of technological progress in environmental-economic models, see Löschel (2002), Carraro and Galeotti (2002).

6.1.2 Decomposition Analysis for Energy Consumption and Energy Intensity in the Manufacturing Sector

To explore the contribution of the various factors to the development of energy consumption over time, a decomposition analysis is applied. Energy consumption at the industry (manufacturing) level can be decomposed using the following identities

$$e_{it} = \frac{E_{it}}{P_{it}} \tag{6.1}$$

where E_{it} is final energy consumption by sub-sector i in period t, P_{it} is the output (value added) by sub-sector i in period t (in constant prices), and e_{it} is the energy intensity of sub-sector i in period t. Total final energy demand in manufacturing in t, E_t, can then be written as

$$E_t = \sum_i E_{it} = \sum_i e_{it} \cdot P_{it} = \sum_i e_{it} \cdot \alpha_{it} P_t \tag{6.2}$$

where $P_t = \sum P_{it}$ is the total output in manufacturing, and $\alpha_{it} = P_{it} / P_t$ is the share of sub-sector i's output in total manufacturing output. Thus, the energy intensity of the manufacturing sector in period t, I_t, can be expressed as

$$I_t = \frac{E_t}{P_t} = \sum_i e_{it} \cdot \alpha_{it} \tag{6.3}$$

Using a Laspeyres-type decomposition, the change in energy intensity in year t compared to a base year 0 can then be decomposed into:

$$I_t - I_0 = \underbrace{\sum_i (e_{it} - e_{i0}) \cdot \alpha_{i0}}_{(a)} + \underbrace{\sum_i (\alpha_{it} - \alpha_{i0}) \cdot e_{i0}}_{(b)} + \underbrace{\text{Residual}(t,0)}_{(c)} \tag{6.4}$$

Thus, the change in *energy intensity* between 2 years can be interpreted as being the result of the following effects:

(a) change in energy intensity at the level of sub-sectors, which captures intra-sectoral structural change as well as actual changes in energy efficiency (efficiency-effect);
(b) change in the shares of the sub-sectors (*inter-sectoral structural change*);
(c) residual term, which may be re-allocated to the other components.[8]

---

[8] The method used in the subsequent analyses is the one proposed by Sun (1998).

Similarly, *total final energy consumption* in the manufacturing sector can be expressed as

$$E_t - E_0 = \sum_i (e_{it} - e_{i0}) \cdot \alpha_{i0} \cdot P_0 + \sum_i (\alpha_{it} - \alpha_{i0}) \cdot e_{i0} \cdot P_0 + \sum_i (P_t - P_0) \cdot e_{i0} \cdot \alpha_{i0} + \text{Residual}(t,0) \tag{6.5}$$

where, in addition to the effects described above,

(d) reflects the change in total output in the manufacturing sector (*activity effect*).

The results of the decomposition analysis for the final energy demand in the West German manufacturing sector (including other mining) between 1970 and 1994 are presented in Fig. 6.1. Thus, the increase in energy consumption resulting from an increase in production has been more than offset by the structural effect and, in particular, by lower energy intensity, i.e. intra-sectoral change and improved energy productivity.

Figure 6.2 shows energy intensity, energy intensity at constant structure, i.e. keeping the same output shares as in 1970, and at constant sub-sector intensities, i.e. keeping the same energy intensities as in 1970. Between 1970 and 1994, actual energy intensity (at constant 1985 prices) dropped from over 5 PJ/billion DM to about 3 PJ/billion DM.

Clearly, without structural change towards less energy-intensive outputs, and without improved energy efficiency, total energy intensity would have been much higher. About three quarters of the observed reduction in total energy intensity can be attributed to the energy-efficiency effect, while the structural-change effect accounts for about one quarter. The development of energy intensity in some energy-intensive sub-sectors is displayed in Fig. 6.3.

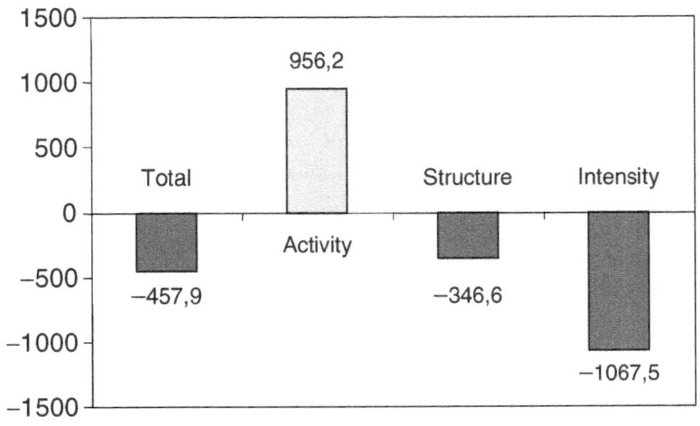

Fig. 6.1 Decomposition of the final energy demand in manufacturing and other mining in West Germany between 1970 and 1994 in PJ (Diekmann et al., 1999)

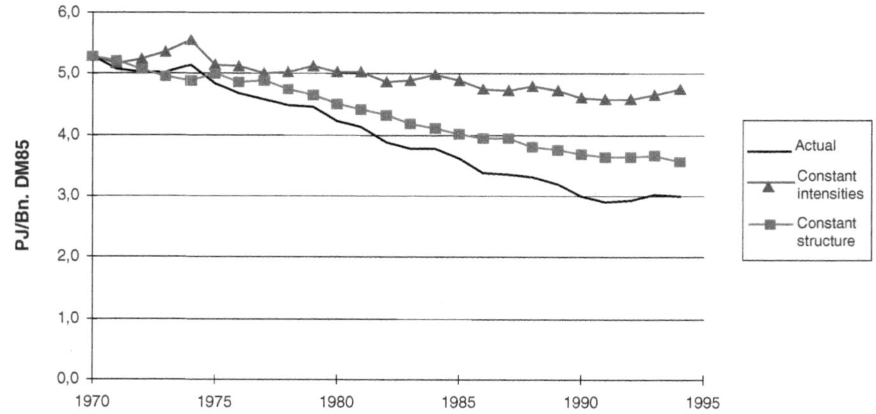

Fig. 6.2 Energy intensity in manufacturing and other mining in West Germany between 1970 and 1994 in PJ/billion DM (1985) (Diekmann et al., 1999)

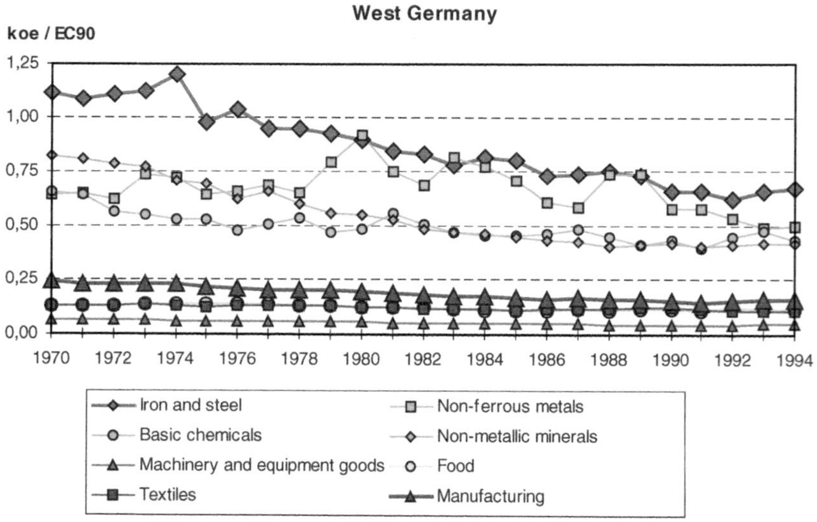

Fig. 6.3 Energy intensity in manufacturing and selected sub-sectors in West Germany between 1970 and 1994 (Odyssee Database)

6.1.3 Estimations of Determinants for the Development of Fuel Intensity

The outcomes of the decomposition analyses can be used to empirically test the impact of various determinants on the development of energy intensity. For fuel intensity, multiple linear regression models were estimated to assess the relevance of fuel prices and other determinants. At the level of total manufacturing, two types of regression equations were estimated for the development of fuel intensity in the West-German manufacturing sector between 1970 and 1994. The first regression equation contains actual fuel intensity as the LHS-variable. To explicitly explore structural change effects, the second equation contains the structural change effect for fuel intensity as the LHS-variable. In general, the following RHS-variables were included in the equations:

- real weighted industrial contemporaneous and lagged fuel prices; the real consumption shares of the fuels serve as weights; the expected sign is negative because of factor substitution and induced technological change;
- contemporaneous and lagged capital investment; investment figures are not limited to energy-specific investments. The working hypothesis is that improved energy efficiency is a by-product of increased productivity, i.e. higher investments result in lower energy intensity; thus, the expected sign is negative;
- capacity utilisation in order to measure the short-term influence of the business cycle; the expected sign is ambiguous: it may be positive since higher capacity utilisation implies the use of less efficient equipment and plants, or negative since higher capital usage implies less energy consumption for idle capacity;
- degree days to measure the impact of temperature variations; the expected sign is negative since a warmer than average climate will require less heating services.

OLS-estimation results for actual fuel intensity and for the structural change effect appear in Tables 6.2 and 6.3, respectively, together with the values of R^2 and the Durbin–Watson test statistic for first-order autocorrelation.[9] Variables were not included in the equations if they had no impact on the estimation results. In the specified regression equations, all variables were entered as growth rates. Thus, the constant term represents a time trend in the levels, and is supposed to capture the effects of autonomous technological change. Given that all the variables are in first differences, the estimated models explain quite a large part of the variation in the dependent variables, as indicated by the (uncorrected) R^2 of more than 0.5.

Estimation results for total fuel intensity indicate that lagged fuel prices and contemporaneous capital investments have a very significant negative impact on fuel intensity in total manufacturing. As shown in Table 6.2, the individual level of significance for the test statistic associated with these variables is below 1%. However, the magnitude of these effects is small compared to the impact of autonomous technological change. Similarly, estimation results for the structural change effect as shown in Table 6.3 indicate that lagged fuel prices and contemporaneous investments have a very significant negative impact on the structural change effect of fuel intensity, while the parameter for capacity use is, as expected, significant

[9] For the estimations presented in this section, extensive misspecification tests were conducted. These comprised more specifically: individual tests for normality (Bera-Jarque), a RESET-type F-test for linearity (in RHS-variables), heteroskedasticity, and autocorrelation (see, for example, Spanos, 1990). The final model specifications presented in this section were selected based on these specification tests together with the usual test statistics (DW, goodness-of-fit, etc.).

Table 6.2 Results for fuel intensity in the manufacturing sector in West Germany (1970–1994)

Variable[a]	Coefficient		Standard error
CONSTANT	−2.646**		0.453
FUEL PRICE (t)	−0.00		0.019
FUEL PRICE (t−1)	−0.061**		0.019
INVESTMENT (t)	−0.162**		0.048
R^2		0.562	
Durbin–Watson		2.052	

**Individually statistically significant at 1% level
[a]All variables enter equation as growth rates

Table 6.3 Results for the structural change effect of fuel intensity in the manufacturing sector in West Germany (1970–1994)

Variable[a]	Coefficient		Standard error
CONSTANT	−0.091		0.564
FUEL PRICE (t)	−0.015		0.023
FUEL PRICE (t−1)	−0.050*		0.024
FUEL PRICE (t−2)	−0.029		0.023
INVESTMENT (t)	−0.188*		0.077
CAPACITY USE (t)	−0.356*		0.173
R^2		0.509	
Durbin–Watson		2.109	

*Individually statistically significant at 10% level
[a]All variables enter equation as growth rates

and positive. These findings are consistent with the view that both higher fuel prices and higher investment rates not only reduce fuel intensity, but also bring about structural changes in favour of less energy-intensive branches. In addition, a booming economy, which is approximated by capacity use, appears to increase the output share of more energy-intensive products. Finally, unlike fuel intensity at the level of manufacturing, autonomous technical change is not statistically significant (and rather small) in the structural change equation. Combining both results would imply that autonomous technical change is likely to play an important role in terms of energy efficiency (and intra-sectoral structural change), at least for some sub-sectors. Energy efficiency is assumed to be incorporated in the measure of energy-intensity at the level of sub-sectors. However, estimations for the energy efficiency effect, i.e. the change in energy intensity at constant structure, did not produce any valid results at the aggregate level of total manufacturing. This suggests that the energy price effects vary considerably across sectors and that the data does not allow identification of a common pattern.[10]

[10]This conclusion is also supported by Miketa (2001), who estimated the impact of energy prices on final energy intensity in 39 countries, but did not find a common pattern across branches. In only six of the ten branches examined were energy prices shown to have a negative and statistically significant impact on energy intensity.

6.2 Econometric Analyses for the Iron and Steel Sector

In this section, the relevance of various potential determinants for the invention, adoption and diffusion of energy-efficient technologies is assessed econometrically in more detail for the steel industry sector in Germany. Besides energy prices, the impact of other determinants such as industry structure, expenditures for research and development or sunk costs associated with incumbent technologies are also considered. The analysis starts with a presentation of the various technological paradigms within the steel industry. Then three types of econometric analysis are performed which also differ by data source and data availability. The first type of regression analysis refers to the aggregate level of fuel intensity in the production of pig iron, for which data is taken from the Odyssee database used earlier. The second and third types of analysis are conducted on a more disaggregate level of the steel sector. They are based on the hypothesis that technical progress and thus improvements in energy performance are primarily realized via new capital goods for putty-clay type production technologies (Dosi, 1988). Silverberg (1988) argues that innovation proceeds through the diffusion of best-practice technologies. Subsequently, the third type of econometric analysis explores the determinants of the energy performance of best-practice technologies.

6.2.1 Technological Paradigms[11]

There are two basic technologies (technological paradigms) for steel making: blast furnace/oxygen (BOF) converters and electric arc furnaces (EAF):[12]

- *oxygen steel production*, i.e. the process of producing primary materials following the route sintering plant (ore concentration)/coking plant – blast furnace (iron making) – converter (steel production), as well as
- *electric arc furnace steel production*, i.e. the process of producing secondary materials principally in electric arc furnaces (to a lesser extent in induction furnaces) based on smelted down scrap.[13]

The subsequent procedural steps consist of *ladle metallurgy* to treat the liquid crude steel (adjustment of the material features and alloy composition) and casting

[11] The description of the production processes in the steel sector is, to a large extent, based on Schleich et al. (2002) and Schön and Ball (2003).

[12] Two other methods used to produce crude steel were not included in the detailed process illustration in the model due to their low significance in the period under observation: the basic Bessemer process has not been operative in Germany since the second half of the 70s and the open hearth process was phased out at the beginning of the 80s and shortly after reunification in the former German Democratic Republic.

[13] Scrap is also used (in small amounts) in oxygen steel production to regulate the temperature of the exothermic conversion process in the converter. Primary materials can also be used when producing electric arc furnace steel. One example, which is practised at only one location in Germany, is the *direct reduction* of iron ore to *sponge iron* (DRI direct reduced iron) using natural gas which is used in the electric arc furnace steel process.

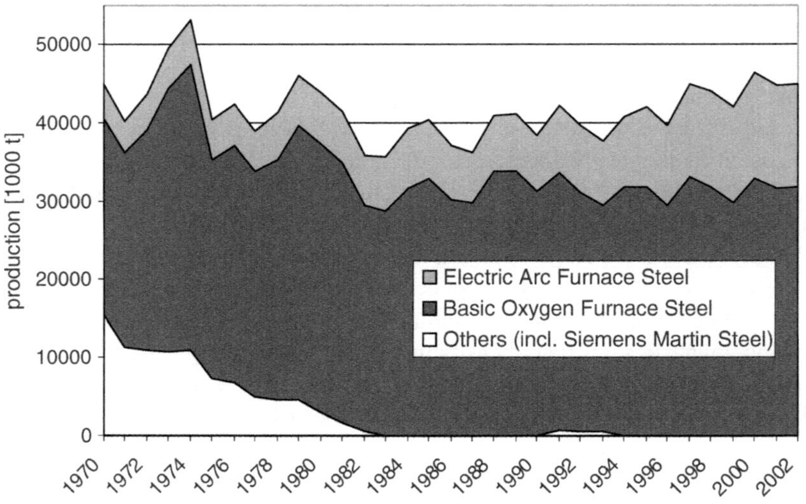

Fig. 6.4 Process lines of crude steel production (Odyssee database, Wirtschaftsvereinigung Stahl/ VDEH, several volumes, own calculations)

and rolling which are excluded from the model-based examination made here. The production of electric arc furnace steel is more attractive from an energetic viewpoint, since it requires less than half the primary energy demand of the blast furnace-oxygen steel route. In Germany, energy costs account for about 25% of total production costs for both technology lines (Ameling, 2004). The development of the production amounts and the shares of the basic technologies in Germany are shown in Fig. 6.4. Whereas the production of oxygen steel has fluctuated around 30M t/a over the last 30 years, the production of electric arc furnace steel has increased continuously from around 4M t/a in 1970 to 6.5M t/a in 1980 to over 13M t/a since then. Apart from a few exceptions in former East Germany, other production technologies like the Siemens–Martin procedure have not been used since the mid 1980s. In the past, the technical development in the steel industry was characterised by concentration on a few larger capacity production plants. The 104 blast furnaces operating in the Federal Republic of Germany in 1970 fell to 80 in 1980 and to 42 in 1990. In 2000 there were 22 left in Germany, of which only 16 were actually being operated. The number of oxygen steel converters decreased from 47 to 26, electric arc furnaces from 71 to 29 between 1980 and 2000.[14]

Figures 6.5 and 6.6 show the time path for fuel intensity and for electricity intensity, respectively, in the German steel sector. Since 1970, fuel intensity has declined

[14] See WV Stahl/VDEH (Wirtschaftsvereinigung Stahl/Verein Deutscher Eisenhüttenleute; Hrsg.): Statistisches Jahrbuch der Stahlindustrie, diverse Jahrgänge. Verlag Stahleisen Düsseldorf; VDEH (Verein Deutscher Eisenhüttenleute).

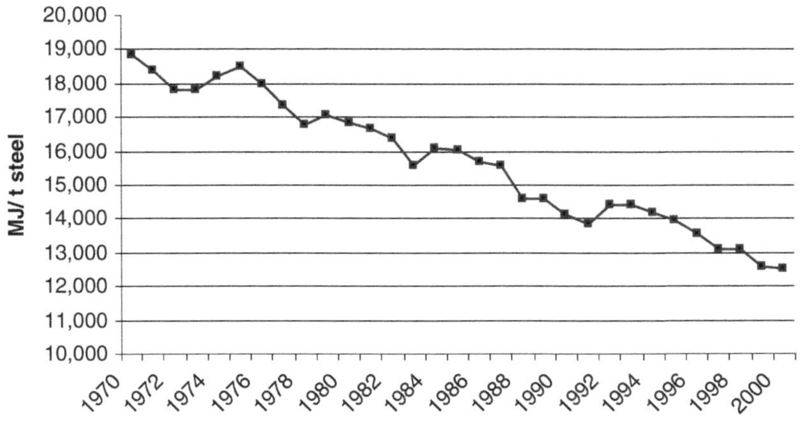

Fig. 6.5 Average specific fuel consumption for steel production in Germany (Odyssee database, Wirtschaftsvereinigung Stahl/VDEH, several volumes, own calculations)

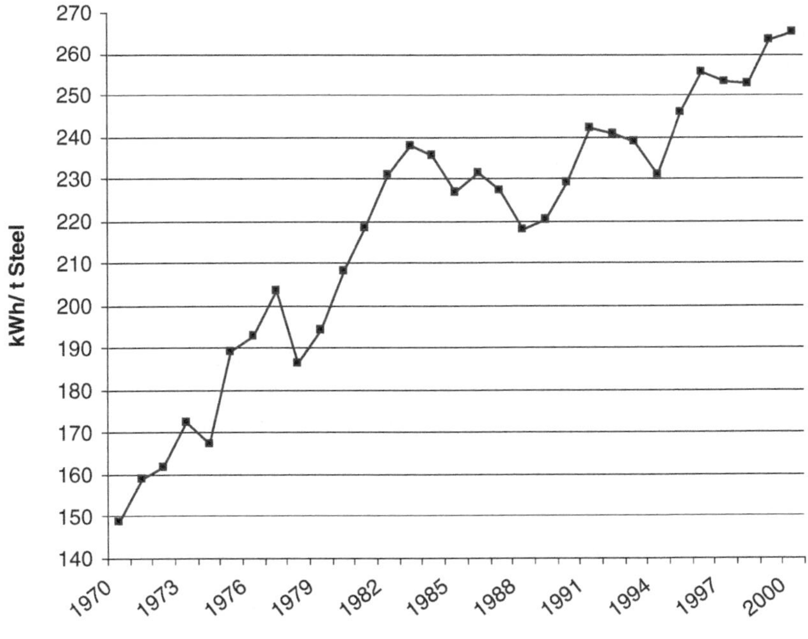

Fig. 6.6 Average specific electricity consumption for steel production in Germany (Odyssee database, Wirtschaftsvereinigung Stahl/VDEH, several volumes, own calculations)

by about one third and electricity intensity has increased by about three quarters. Clearly, these figures reflect both a switch from fuel-based BOF to electricity-based EAF as well as technological progress in terms of energy efficiency.

In the period under review, the technical measures to improve energy efficiency which are reflected in the chronological changes of the energy input structures include, among others, decreasing the consumption of the reducing agent in iron making – e.g. by partly substituting coke with injected pit coal, fuel oil or scrap plastics – measures in integrated ironworks, coking plants and sintering plants as well as control technology measures and the optimisation of the energy supply in electric arc furnace steel works.[15]

6.2.2 *Estimations of Aggregate Fuel Intensity in the Production of Pig Iron*

In the first equation, where data is based on the Odyssee database, fuel intensity in the production of pig iron (as measured in PJ/ billion €) is regressed on the set of RHS-variables defined in Sect. 5.2.3.[16] Thus, this analysis is similar to the one carried out earlier on a more aggregate level. The only other similar study known is Miketa (2001) who regressed energy intensity (as measured by final energy consumption to real output) in the industry sectors of 39 countries on contemporaneous output, investment, and energy prices. For the iron and steel sector, Miketa (2001) found output and energy prices to be statistically significant at the 5% and 10% level, respectively.

Between 1970 and 1994, fuel intensity in the production of pig iron in West Germany dropped by over 30%. At the same time, real fuel prices increased by about 73%. Estimation results for the multivariate case are presented in Table 6.4. The findings indicate that fuel prices of lag two (actually growth rates) are significant well below the 10% level, while fuel prices of lag one are only significant above a 16% level, but show the expected negative sign. Investment and capacity use are not statistically significant at any reasonable level.[17] Similarly, the variable degree days are not statistically significant either. Not surprisingly, in contrast to the housing or the tertiary sector, thermal energy use for room heating purposes in the steel

[15] A comprehensive energetic evaluation of oxygen steel production and its CO_2 emissions is given by Aichinger et al. (2001). A detailed examination of the electrical energy consumption of electric arc furnace steel production (arc furnaces) can be found in Köhle (1992) and Köhle (1999).

[16] The regression equations in Sects. 5.2.3 and 5.3.2 were estimated individually.

[17] Although the level of aggregation, the countries considered, and the variables included are quite different in Miketa (2001) than in the analysis presented here, investment also turned out to be insignificant and negatively correlated with energy intensity.

Table 6.4 Results for fuel intensity in the production of pig iron in West Germany (1970–1994)

Variable[a]	Coefficient	Standard error
CONSTANT	−0.447	0.812
FUEL PRICE (t–1)	−0.056	0.039
FUEL PRICE (t–2)	−0.086*	0.042
INVESTMENT (t)	−0.032	0.030
CAPACITY USE (t)	−0.070	0.085
DEGREE DAYS (t)	−0.170	0.103
R^2	0.271	
Durbin–Watson	2.055	

*Individually statistically significant at 10% level
[a]All variables enter equation as growth rates

sector cannot be expected to play a major role for energy intensity. The fact that only one variable turned out to be statistically significant, may be due to the relatively low degrees of freedom: there are only observations from 25 years available to estimate six variables. Likewise, there may be collinearity among the RHS variables.[18]

6.2.3 *Estimations for the Diffusion of the Electric Arc Furnace*[19]

The level of analysis is quite aggregated in the previous section. More specifically, the processes which map the changes in energy prices and other factors into changes in energy intensity are not explicit. Likewise, the previous analyses do not allow the effects of price changes on factor substitution to be distinguished from the effects on the price-induced adoption and diffusion of new technologies. In terms of innovation, the steel sector may best be characterized as "supplier dominated". This type of sector contributes relatively little to innovation.[20] Instead, technical progress, which, in the steel sector, is mainly process-integrated, is primarily realized via new capital goods. Dosi (1988) and also Silverberg (1988) stress that, in such sectors,

[18]While collinearity leads to "inflated" estimates for the parameter variances and reduces the level of significance for individual parameter estimates, the parameter estimates are still unbiased. The best "cure" to treat collinearity is to collect more data, which – because of limited data availability for West-Germany following the German reunification – is not an option.

[19]Sections 6.2.3 and 6.2.4 represent a shorter and slightly modified version of Schleich (2007). Helpful comments by two anonymous reviewers, and by Carsten Nathani, Inga Fischer, Michael Schön, Rainer Walz and Katja Schumacher are gratefully acknowledged.

[20]See Pavitt (1984). Nevertheless, the "supply sectors" usually develop new technologies in close co-operation with the "demand sectors". Often, the demand sectors have their own research institutions available at national and international level, as is the case for the steel industry.

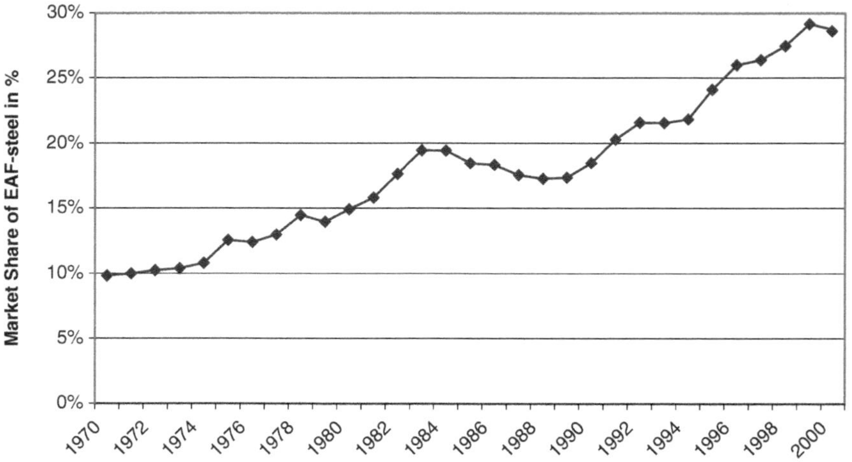

Fig. 6.7 Diffusion of EAF-steel in Germany since 1970 (Wirtschaftsvereinigung Stahl/VDEH, several volumes)

innovation proceeds through the adoption and diffusion of best-practice technologies. Thus, technological progress is incorporated in the capital goods. In the steel sector – as in most other energy-intensive sectors – technology substantially limits factor substitution. Technologies are best described as "putty-clay"; companies may chose from among different technologies when making their investment decisions, but thereafter the input structure is basically fixed. Thus, the observed changes in energy intensity in many energy-intensive sectors are primarily the result of the adoption and diffusion of new technologies. In the case of the steel industry, the observed changes in energy intensity may be the result of

- a switch in technological paradigms within the steel sector; the "diffusion curve" for EAF-steel is displayed in Fig. 6.7;
- a change in the energy performance of new technologies within technological paradigms over time.

In this section, an analysis is made of the impact of energy and other input prices on the choice of technologies in the steel sector in Germany. The approach taken is similar to Reppelin-Hill (1999), who explored the (linear) relationship between several explanatory variables and the diffusion of EAF in 30 steel-producing countries over 25 years.[21] Using a random-effects model, she finds that neither electricity

[21] In general, imposing a linear relationship on the data might be a problem if the actual development of the shares followed an S-shaped curve and if the observed values were near the saturation point for the EAF technology. However, Fig. 6.4 suggests that for Germany this is not the case. In addition, comparing the current share of EAF-steel in Germany of about 27% with the share of EAF-steel in the US of about 43% (Ruth and Amato, 2002, p. 548) suggests that the EAF-share has not yet reached its limit in Germany.

nor coking coal prices enhance the diffusion of EAF-steel. In this section, a similar econometric analysis assessing the determinants for the diffusion of EAF steel in Germany is conducted for a longer period (1970–2000).[22] The LHS-variable is the share of EAF-steel in total crude steel production, EAFSHARE.[23] Besides the lagged EAFSHARE,[24] the set of RHS variables consists of:

- prices of the main inputs in the production of EAF-steel: scrap (PSCRAP) in €/kt and electricity in €/kWh (PPOWER); the expected sign of the parameters associated with these input prices is negative;
- price of coking coal in €/kt (PCOAL), and the price of iron ore in €/kt (PIRON), which comprise the main inputs in the production of basic oxygen or open hearth furnace steel; both variables are expected to be positively correlated with the share of EAF-steel;[25]
- the share of accumulated BOF-steel production to accumulated EAF-steel production (CUMSHA); to calculate CUMSHA the total tons of steel produced by either technology since the beginning of the study period were used; CUMSHA is included as a proxy for capital obsolescence or path dependence. The expected sign is negative because these sunk costs are considered to be a barrier to the diffusion of the newer EAF-technology.

All prices enter the regression as real values based on the year 1995. OLS-estimation results are shown in Table 6.5. The findings indicate that all variables exhibit the expected signs and are – except for the price of iron (PIRON) – individually statistically significant at the 10% level. Thus, in contrast to Reppelin-Hill (1999), not only the price for scrap, but also the prices of the energy carriers electricity and coking coal turn out to be statistically significant. Thus, material and energy input prices appear to play a significant role in the decision to produce with the EAF technology. In addition, the findings for CUMSHA support the hypothesis that path dependencies stemming from high fixed costs dampen the diffusion of EAF-steel.

6.2.4 Estimations for the Development of Energy Efficiency

In this section, econometric analyses are carried out to assess the empirical relevance of several determinants of energy efficiency within the technological paradigms for

[22] Reppelin-Hill (1999) also focuses on the impact of trade openness on the diffusion of EAF-steel. Since this aspect is more relevant in an international analysis, it is neglected in the analysis conducted here.

[23] Lack of data precluded the use of capacity, which may be more appropriate in the light of the high fixed capital associated with the production of steel.

[24] Note that the lagged values of EAFSHARE also capture the possible impact of lagged input prices on the current share of EAF-steel production.

[25] Of course, BOF-steel and EAF-steel are not perfect substitutes. Differences in the quality imply that electric arc furnace steel is mainly used for "long products", whereas the manufacturing of "flat products" (sheets) remains primarily a domain of oxygen steel.

Table 6.5 Estimation results for the share of electric arc furnace steel (EAFSHARE) in crude steel production (1970–2000)

Variable	Coefficient	Standard error
CONSTANT	−0.447	0.147
EAFSHARE (t−1)	−0.056	0.237
PSCRAP	−0.086*	0.081
PPOWER		0.198
PIRON	−0.032	0.446
PCOAL	−0.070	0.228
CUMSHA	−0.170	0.021
R^2 (corrected)	0.97	

Since the model includes the lagged dependent variable on the RHS, the standard Durbin–Watson test statistic cannot be applied to test for serial correlation. Instead Durbin's alternative test (Maddala, 1992) was conducted. Based on this test, serial correlation in the error terms does not appear to be a problem

*Individually statistically significant at 10% level

**Individually statistically significant at 5% level

steel production. For BOF-steel, the specific fuel consumption of the best-practice BOF technology is explored. Similarly, for EAF-steel, the specific electricity use of the best-practice EAF technology is analysed. In both cases, data were taken from Schleich et al. (2002) and are based on observed (i.e. adopted) technologies on the market over the time horizon 1980–2000. More specifically, the data stem from technology-based bottom-up analyses and were discussed with experts from the German steel association (VDEH). Of course, given the relatively short period, the interpretation of the results is subject to the limited number of degrees of freedom.

The representation of technological progress as a smooth process in the econometric model appears justified. For example, Fri (2003) and Arentsen et al. (2002) argue that technological change in the energy sector is realized over long periods because of the large number of actors and corresponding actions, decisions and experiences involved. These are affected by numerous aspects including social, institutional, political, managerial, technological or financial dimensions. As a result of this complex set of relations, technological change emerges as an incremental process. The determinants considered reflect factors affecting the costs and benefits of the new technologies to the adopters as well as factors affecting the technical development of energy efficiency. Thus, the determinants generally include relative input prices and R&D expenditure by the steel industry and by the mechanical engineering and electrical engineering sectors. In addition, indices were included which represent industry concentration in the production of steel in Germany. From a theoretical point of view, the impact of firm size or industry concentration on the adoption of new technologies is ambiguous.[26] On the one hand, large firms or firms with a larger market share may

[26] See Hall and Kahn (2003, p. 9) or Hall (2004, p. 22) for brief recent overviews and for some empirical evidence.

use their market power to appropriate the costs associated with the adoption of new technologies. Such up-front costs not only include investment costs, but also training workers, marketing, or expenses for research and development. Similarly larger firms are more likely to have internal financial resources available, and to have better access to capital markets for financing the adoption of new technologies. In addition, larger companies may be able to spread the potential risks associated with the adoption of new technologies better because they tend to be more diversified in terms of the technologies installed. They are more likely to be in a position to test new technologies while maintaining operation of the old ones as a safety cushion (at lower production levels). Finally, larger firms may capture the economies-of-scale effects associated with the implementation of new technologies faster and they can spread the fixed costs of adoption across more production units. On the other hand, larger firms may also be more bureaucratic and suffer from so-called X-inefficiencies. Such inefficiencies may be the result of complex and time-consuming decision processes, or of internal agency-related problems, such as lack of observability of individual behaviour. Similarly, the degree of concentration of new technology providers may also have an impact on innovation and diffusion. Since highly concentrated providers tend to charge higher prices, they may slow diffusion, but they may also be in a better position to determine a common standard which increases the benefits of adoption.[27] Despite these considerations it was not possible to explore this aspect because there is no data available on the industry concentration of suppliers to the steel sector.

6.2.4.1 Estimations for the Development of Best-Practice Fuel Consumption for BOF-Steel

To estimate the fuel consumption of the best-practice trajectory for BOF, specific fuel use (in GJ/t steel) is used as the LHS variable.[28] The set of RHS variables consists of:

- The relative price of coal (€/kt) to the price of steel (in €/kt) (RELPCS) and the relative price of iron ore (€/kt) to the price of steel (in €/kt) (RELPIS). To allow for some lead time, both variables are entered with a lag of two (t-2). According to a cost- or profit-pressure hypothesis, they are supposed to capture the impact of input prices on the invention, the adoption and the diffusion of more fuel-efficient best practice technologies in BOF steel production. The expected sign of the parameters associated with these relative prices is therefore negative.
- The sum of public and industry R&D expenditure for the steel industry (RDSTL) (in 1995 billion €). R&D efforts in the steel sector are expected to result in the development of new technologies which – since the steel sector is quite energy-intensive – are also more energy-efficient. To allow for some lead time, the average

---

[27] Hall (2004, p. 21).

[28] A similar approach is used, for example, by Lutz et al. (2004).

of the R&D expenses in the steel industry in t and t-1 are used in the concrete model specification. Since greater research efforts should lead to better energy performance, the sign of the parameter estimate associated with RDSTL is expected to be negative.

- The sum of R&D expenditure in the electrical engineering (EE) and mechanical engineering (ME) industry sectors (RDEEME) (in 1995 billion €). To allow for some lead time, an average of the contemporaneous and lagged (t-1) R&D spending in these sectors is used.[29] Since an increase in the general spending level in EE and ME is also expected to improve the energy efficiency of new technologies in the steel industry, the expected sign is negative.
- The six firm concentration index (CR)[30] to assess the impact of industry concentration in the steel industry on the adoption and diffusion of best-practice technologies in the steel industry.

The OLS-estimation results are shown in Table 6.6.

In Table 6.6, all signs are as expected, and, except for the relative price of iron ore,[31] all parameters turn out to be statistically significant at least at the 10% level, despite the rather small size of the sample. Thus, in general, the results are consistent with the profit/cost-pressure hypothesis. As for the magnitude, the estimation results suggest that the price effects are relatively small. For example, an increase in the relative price of coking coal to steel by 10% reduces specific fuel consumption of the best practice BOF-technology by about 0.1%.[32] Not surprisingly, R&D expenditure in the steel sector has a greater effect (by almost a factor of 10) than a general increase in the R&D spending in the mechanical engineering and electrical engineering sectors. Also note that the lag structure of input prices and R&D expenditure is consistent with the view that higher input prices trigger R&D measures which, in turn, result in the development of more energy-efficient technologies.[33] Finally, an increase in the concentration of the industry as measured by the six-firm concentration index turns out to have a small, but statistically significant, negative impact on best-practice fuel demand. Overall, the estimated regression equation explains a fairly high share of the variation in the best-practice specific fuel consumption,

---

[29] Contemporaneous R&D expenditures capture the fact that innovations are often the outcome of joint efforts between developers and users. Thus, R&D efforts in the mechanical and electrical engineering sectors which lead to reductions in energy consumption in the steel sector in the same period would be accounted for.

[30] Sales share of six largest firms (in terms of sales).

[31] The relative price of iron ore would be statistically significant at the 15% level, however.

[32] To calculate this number, average sample values for specific fuel consumption and for the relative price of coking coal to steel were used together with the parameter estimate associated with RELPCS (t-2).

[33] Note that such a correlation between input prices and R&D expenditure may result in high variances, but not in biased parameter estimates.

Table 6.6 Estimation results for the best-practice fuel use in BOF steel production (1980–2000)

Variable	Coefficient		Standard error
CONSTANT	19.87**		0.116
RELPCS (t–2)	–0.576*		0.193
RELPIS (t–2)	–0.461*		0.301
RDSTL (t, t–1)	–2.031**		0.109
RDEEME (t, t–1)	–0.248**		0.003
CR	0.003*		0.001
R^2 (corrected)		0.99	
Durbin–Watson		1.88	

*Individually statistically significant at 10% level
**Individually statistically significant at 1% level

as was also the case in Lutz et al. (2005). Further, the value for the Durbin–Watson statistic dos not hint at autocorrelation in the error terms.

6.2.4.2 Estimations for the Development of Best-Practice Electricity Consumption for EAF-Steel

In the regression equation for the electricity consumption of the best-practice trajectory for EAF, specific electricity use (in MJ/t steel) is used as the LHS variable. The set of RHS variables is similar to the set of variables for estimating specific fuel use for the BOF best-practice trajectory. Only the variables reflecting the energy costs and the material input costs differ. The cost variables included are:

- The relative price of electricity (€/MWh) to the price of steel (in €/kt) (RELPPS) and the relative price of scrap (€/kt) to the price of steel (in €/kt) (RELPSCS). As was the case for the best-practice BOF technology, a lag of 2 years produced the best estimation results. The profit/cost-pressure hypothesis suggests that the signs of the parameters associated with these relative prices should be negative.

Table 6.7 presents the OLS-estimation results for specific electricity use of the best-practice EAF-technology.

As can be seen in Table 6.7, the parameter estimates associated with all price and expenditure variables are statistically significant at least at the 10% level. Unlike the fuel use of the best-practice BOF-technology, the concentration index for the electricity use of the best-practice EAF technology is far from becoming statistically significant in the estimated regression equation. Again, the impact of R&D expenditure in the steel sector is about ten times greater than the impact of R&D spending in the electrical engineering and mechanical engineering sectors. Once more, the estimated regression equation explains a high share of the variation in the best-practice specific electricity consumption.

Table 6.7 Estimation results for the best-practice electricity use in EAF steel production (1980–2000)

Variable	Coefficient		Standard error
CONSTANT	2.205**		0.021
RELPPS (t–2)	–0.066*		0.024
RELPSCS (t–2)	–0.026*		0.013
RDSTL (t, t–1)	–0.266**		0.017
RDEEME (t, t–1)	–0.027**		0.001
CR	–0.001		0.000
R^2 (corrected)		0.99	
Durbin–Watson		1.62	

*Individually statistically significant at 10% level
**Individually statistically significant at 1% level

6.3 Conclusions

The econometric analysis of the determinants underlying the observed decline in total fuel intensity in the West German manufacturing sector imply that higher fuel prices not only result in lower fuel intensity, but also trigger structural changes towards less energy-intensive products. Results of the case study on energy use over time in the iron and steel sector in Germany first suggest that it is important to account for intra-sectoral structural change such as a switch from basic oxygen steel production to electric arc furnace steel production when analysing energy use and emission trends in this sector. The econometric analyses indicate that material and energy input prices play a significant role in this switch in technological paradigms. In addition, path dependencies stemming from high fixed costs dampen the diffusion of EAF-steel. In addition, for both paradigms technological change in terms of improved energy use is found to be positively related to higher energy prices, to higher R&D efforts in the mechanical engineering sector, but is dampened by concentration in the steel industry.

Thus, the empirical evidence presented in this chapter supports the general view that for energy-intensive industry price policies in particular are likely to be effective climate policy instruments, which provide incentives for the invention, adoption and diffusion of energy-efficient technologies in sectors with a high energy cost share. As such, the results are generally consistent with the hypothesis of induced technological change.

Chapter 7
Barriers to the Diffusion of Energy Efficiency in the German Commercial and Services Sectors[1]

In this section, econometric methods are applied to empirically assess the relevance of various factors of influence on energy-efficiency barriers in the German commerce and services sectors (small commercial businesses and private and public service organisations).[2] For various barriers, a separate regression equation is estimated econometrically. In this sense, the analyses conducted are similar to DeGroot et al. (2001). The barriers include, for example, lack of information about energy-consumption patterns or energy-efficient measures. Based on the estimation results, the statistical significance of various "explanatory" variables such as energy consumption, size of the company, or differences across sub-sectors can be assessed. Moreover, it is examined to which extent energy audits[3] help overcome these barriers, as has often been suggested (Enquête Commission, 1989; Geiger et al., 1999; European Commission, 2006). The estimation results presented in this chapter make it possible to evaluate the effectiveness of energy audits conducted by utilities, engineering firms, or industry sector associations.

The next subsections describe the survey and the data used, discuss the barriers which are entered into the analyses as "dependent" variables, portray the "explanatory" variables available and present estimation results. The section concludes with policy implications.

---

[1] This chapter is a shorter and revised version of Schleich (2004). Insightful suggestions by Edelgard Gruber and Manuel Frondel are gratefully acknowledged.

[2] In the German energy balances, final energy consumption is partitioned into four end-use sectors: industry, private households, transportation and the combined sector commerce and services.

[3] The term "audit", as used in this chapter, not only includes formal energy audits, but also more general consultations on energy efficiency or the tariff structure. In a formal energy audit, energy consumption in the organisation is analysed and evaluated, measures to reduce energy use and energy costs are identified and assessed and energy use patterns and proposed measures may be reported (European Commission, 2001). Energy audits serve as a tool for energy management and may also form part of an organisation's environmental management system such as EMAS.

R. Walz and J. Schleich. *The Economics of Climate Change Policies.*
Sustainability and Innovation,

7.1 The Data

The required data are taken from a recent representative survey of energy consumption in the commercial and services sector in Germany (Geiger et al., 1999). Table 7.1 displays in more detail the sub-sectors included in the German commerical and services sector, which accounts for about 16% of final energy consumption (FMEL, 2002) and 20% of CO_2 emissions in Germany (FEA, 2002). Unlike two earlier similar surveys which were conducted in 1978 and 1982, the latest survey did not just contain questions on the economic and technical factors of energy use. Questions were also included about energy management, measures taken and obstacles to energy-efficiency improvements.

Because of its heterogeneity, the sector was broken down into several more or less homogeneous divisions that reflect the sub-sectoral structure in official statistics, and where necessary, even in subdivisions. Based on literature research, discussions with specialists, in-depth interviews and plant inspections, a structured questionnaire was elaborated, and almost 3,000 managers of companies and public institutions were interviewed personally by well-trained staff. The sample was based on a quota method: a minimum number of respondents were required in each division in three different groups of company size. Unfortunately, due to a lack of

Table 7.1 Overview of sub-sectors

Sector	No. of observations	Average no. of employees	Average annual energy consumption (MWh)
Metal industry	116	9	125
Car repair industry	78	8	339
Wood working and processing	94	8	209
Bakeries	88	9	378
Butchers	76	8	151
Laundries and dry cleaners	67	19	1,978
Building (interior construction)	98	15	89
Retail trade	296	29	459
Wholesale trade	168	37	837
Banks, insurance companies	129	162	1,622
Hotel industry	129	17	589
Gastronomy	103	10	216
Services[a]	78	10	48
Non-commercial organisations	127	29	508
Public administrations	96	69	934
Hospitals	79	300	8,834

[a]Split for lawyers, architects, small private health services, private agencies, etc.

sufficient data and data compatibility, not all splits could be included in the econometric analysis. Similarly, for the analyses in this section, a full set of observations was only available for about 1,800 organisations.

With regard to energy-efficiency measures, the respondents received a list of sector-specific measures and were asked which had been implemented in their organisations. These lists differed across sub-sectors and referred to the specific production equipment in the industrial sectors and to the building and the heating system in the remaining branches which are characterised by offices. In the sub-sectors dominated by space heating, the lists included technical issues such as insulation of walls and windows, control systems for heating and lighting, heat recovery, as well as organisational measures such as energy analysis, and the establishment of an energy management. In the industrial splits, additional production-oriented measures were included, such as leakage removal in compressed-air systems or investments in energy-efficient cooling. In addition, the survey also asked respondents to judge the relevance of potential energy-efficiency barriers within their organisation, whether they had previously conducted an energy audit, and who had carried out the audit: utilities, engineering firms, industry sector associations, or others.

7.2 Barriers to the Diffusion of Energy Efficiency

Improving energy efficiency is often seen as the fastest and most cost-effective way to achieve a sustainable energy system (e.g. IPCC, 2007). Consequently, strategies for obtaining more energy services such as heat, light or mobility with the same or less energy input have recently attracted increased attention from policymakers and academics alike. Engineering-economic analyses indicate that a significant potential of measures exists, for which the monetary benefits of energy saved exceed additional capital, operating and maintenance costs, but which have not been realised. According to the IPCC Third Assessment Report (IPCC, 2001), about half the technological potential for greenhouse gas emission reductions world-wide is also profitable (IPCC, 2001, p. 174; Synthesis report, Chap. 3, Executive summary). In terms of greenhouse gas emissions, the IPCC (2001) estimates this so-called no-regret potential to range between 10% and 20% of global emissions in the year 2020. According to the UNDP/WEC/DESA (2000), the no-regret potential for all sectors is even higher, at about 20–30%. For the commercial sector in particular, the profitable savings potential in buildings is estimated at 10–20% for the year 2010 and at 30% for 2020. More recently, the Spring European Council Presidency Conclusions stress the need "to increase energy efficiency in the EU so as to achieve the objective of saving 20% of the EUs energy consumption compared to projections for 2020, as estimated by the Commission in its Green Paper on Energy Efficiency (CEU, 2007, p. 20)". In view of that, the Action Plan (European Commission, 2006) outlines a framework of policies and measures for all end-use sectors (residential, tertiary, industry and transportation) and the transformation sector to improve energy efficiency. Accordingly, "additional investment expenditure in more efficient and innovative technologies will be remunerated by the more than €100 billions annual fuel savings" in the EU (European Commission, 2006, p. 3). The commercial buildings (tertiary) sector is estimated to exhibit the highest relative potential for energy savings of 30%. Proposed measures to realise these potentials include implementing energy management systems, promoting public-private energy efficiency funds or financing packages and energy audits in small and medium sized companies and in the public sector. In particular, such policy measures are supposed to help overcome the so-called barriers to energy efficiency which are preventing energy-efficiency measures from being realised. Overcoming these barriers would contribute significantly to achieving international greenhouse gas emission targets and reducing the reliance on imports of fossil fuels at low cost.

A clear comprehension of the nature of these barriers is decisive when designing cost-efficient policy measures. So far, most empirical analyses on barriers to energy efficiency are in the form of case studies, where theory-based hypotheses are derived from various (partially overlapping) concepts grounded in neo-classical economics, institutional economics, organisational theory, sociology, and psychology (DeCanio, 1994; de Almeida, 1998; InterSEE, 1998; Ramesohl, 1998; Schleich et al., 2001a; Sorrell, 2003; Sorrell et al., 2004). Complementary to the case study approach, a few analyses exist which rely on surveys to explore the empirical relevance of barriers to energy efficiency, including Brechling and Smith (1994), Scott (1997), DeCanio (1998), De Groot et al. (2001) and Schleich and Gruber (2008).

Setting matters of profitability of the abatement measures aside, in the following section, some of the obstacles identified in the "barriers literature" will be explored econometrically, while in this section, variables will be presented that are included in the analyses to reflect those barriers. All "dependent" variables which are assumed to reflect an energy-efficiency barrier enter the regression equations as dummy variables. They take on the value of 1 if – according to the subjective judgement of the interviewee – the stated barrier was relevant in her or his organisation. Otherwise, the value is 0. For some barriers, objective information is available and used here instead of subjective judgements. In total, six types of energy-efficiency barriers are investigated: lack of time, lack of information about energy consumption patterns, lack of information about energy efficiency measures, investment priorities, uncertainty about future energy costs and split incentives.

7.2.1 *Lack of Time*

In companies originating from energy-intensive industries like the power or the iron and steel industries, energy performance affects the core production process and the energy cost share is typically rather high. So, there are strong economic incentives for these companies to find and realise efficiency potentials. By contrast, in the commerce and services sector, the energy cost share is usually low, and investments in energy efficiency do not affect the core production processes. In addition, since companies in these sectors are usually rather small, the indirect or hidden costs associated with investments in energy efficiency, such as overhead costs for energy management, and costs for training personnel, are more likely to be prohibitive. The same may hold for transaction costs which generally include the costs of gathering, assessing and applying information on energy savings potentials and measures, as well as costs to locate efficiency improvement potentials and negotiate contracts with potential suppliers, consultants or installers, or the costs of making, monitoring and enforcing contracts (Coase, 1991). Thus, lack of time to analyse energy efficiency potentials, is likely to constitute a barrier to energy efficiency in the commercial and services sector. The dependent variable TIME, capturing all these aspects, takes on the value of 1 if survey respondents consider lack of time to be a relevant barrier in their organisation.

7.2.2 Lack of Information About Energy Consumption Patterns

Measuring and controlling energy consumption at a disaggregated level is costly to organisations. Labour costs for metering and data management, and investment costs for the metering devices may prevent organisations from installing the appropriate equipment. However, if energy consumption, and hence, energy costs, are not known in detail, the profitability of energy saving measures cannot be properly assessed. The dependent variable EKNOW assumes the value of 1 if the split of final energy consumption into thermal energy and electricity consumption is unknown.

7.2.3 Lack of Information About Measures

There may be a lack of information about energy efficiency measures in organisations for several reasons. First, organisations with a low energy cost share have little incentive to overcome transaction costs and spend resources to find out about new energy savings technologies. Second, information about the performance of energy efficiency measures is a typical public good. Thus, if the production of this public good is left to the private market, "too little" information about energy efficient technologies will be produced. The respective dependent variable INFO has the value of 1 if survey respondents consider lack of information about energy efficient measures to be a relevant barrier in their organisation.

7.2.4 Investment Priorities

A crucial criterion for organisations' investments in energy efficiency is profitability, or the payback period. Both depend on the capital costs for the organisation. Restricted access to capital markets is often considered to be an important barrier to investing in energy efficiency. That is, investments may not be profitable because companies face a high price for capital. As a result, only investments yielding an expected return that exceeds this (high) rate will be realised. Since the price of capital also reflects the risk associated with the borrower, small- and medium-sized companies often have to pay higher-than-average interest rates. Possible explanations include smaller companies' limited ability to offer collateral or potential lenders having to bear higher costs to assess the credit-worthiness of small- and medium-sized companies. When access to the capital market is constrained, the allocation of funds within an organisation becomes even more important. Internal decision making and priority setting will not only depend on hard investment criteria such as rate of return or the payback time of an investment project, but also on soft factors such as the status of energy efficiency, reputation, or the influence of those responsible for energy management within the organisation (Morgan, 1985; DeCanio, 1994). The dependent variable PRIORITY assumes the value of 1 if investment priorities are considered to be a barrier to energy efficiency in the organisation.

7.2.5 Uncertainty About Future Energy Costs

Investing in a more energy-efficient technology may turn out to be unprofitable if energy prices fall after the new technology has been implemented. Hence, there is an option value associated with postponing investments (McDonald and Siegel, 1986; Dixit and Pindyck, 1994) and postponing irreversible investments in energy efficiency may be optimal if future energy prices are uncertain, even though the expected value remains unchanged (Hasset and Metcalf, 1993; van Soest and Bulte, 2001). In addition, since the interviews were conducted in 1997, thus prior to the liberalisation of energy markets in Germany (1998 for electricity, and 2000 for gas), organisations may have (correctly) expected energy prices to fall in the wake of the liberalised energy markets, rendering investments in energy efficiency less profitable. On the other hand, such investments in energy-efficient technologies reduce companies' "risk exposure" to variability in energy prices. The relative magnitudes of these countervailing effects are company-specific and generally ambiguous. Whether energy price volatility increases or decreases incentives to invest in energy efficiency depends – among other things – on a company's attitude towards risk, the expected energy costs and on the irreversibility of the investments (see Howarth and Sanstad, 1995; Ben-David et al., 2000). The dependent variable UNCERT takes on the value of 1 if uncertainty about future energy costs is considered to be an energy-efficiency barrier in the organisation.

7.2.6 Landlord/Tenant Dilemma (Split Incentives)

If a company is renting buildings or office space, neither the landlord nor the company (tenant) may have an incentive to invest in energy efficiency, because the investor cannot appropriate the energy cost savings. The landlord will not invest in energy efficiency if the investment costs cannot be passed on to the tenant, who will benefit from the investment through lower energy costs. On the other hand, the tenant will not invest if he/she is likely to move out before fully benefiting from the energy cost savings. The respective dependent variable RENTED assumes the value of 1 if rented space is considered to be an energy-efficiency barrier in the organisation.

7.3 Determinants

Based on the data provided in the survey, the following "independent" or "explanatory" variables are included as regressors in the equation: energy consumption, size of the organisation, energy audits and dummies for individual sub-sectors.

7.3.1 Energy Consumption

Clearly, organisations' incentives to spend resources to overcome barriers to energy efficiency depend on the energy cost savings expected. Thus, total annual specific energy consumption, ENERGY, is included to reflect the importance of energy consumption and energy costs to the organisation. To control for size effects, specific measures, rather than absolute levels of fuel consumption are used: To create ENERGY, total annual fuel and electricity consumption were added together and divided by the number of employees. Since ENERGY, which is ultimately constructed by taking the natural logarithm of specific energy consumption, is expected to have a negative impact on energy-efficiency barriers, the expected sign for the parameter estimate associated with ENERGY is negative.

7.3.2 Size

Larger organisations are more apt than smaller ones to deal with barriers such as information and other transaction costs, credit constraints, or uncertainty. Thus, the variable SIZE which stands for the number of employees in the organisation is expected to have a negative effect on barriers.

7.3.3 Energy Audits

In the survey, companies were also asked whether they had recently had an energy audit. If the audit was carried out by a utility, an industry branch association, an engineering firm, or others, the dummy variables UTILITY, ASSOC, ENGIN or

Table 7.2 Percentage shares of audits conducted by various types of external consultants

Sector	UTILITY	ASSOC	ENGIN	OTHER	Total
Metal industry	0.06	0.02	0.04	0.02	0.15
Car repair industry	0.08	0.06	0.08	0.01	0.23
Wood working and processing	0.11	0.04	0.07	0.01	0.23
Bakeries	0.13	0.05	0.03	0.02	0.23
Butchers	0.17	0.08	0.09	0.04	0.38
Laundries and dry cleaners	0.12	0.06	0.15	0.00	0.33
Building (interior construction)	0.07	0.03	0.02	0.00	0.12
Retail trade	0.10	0.05	0.08	0.02	0.25
Wholesale trade	0.11	0.10	0.15	0.02	0.38
Banks, insurance companies	0.11	0.05	0.12	0.02	0.29
Hotel industry	0.11	0.04	0.09	0.04	0.27
Gastronomy	0.11	0.02	0.06	0.02	0.20
Services[a]	0.04	0.01	0.06	0.00	0.12
Non-commercial organisations	0.10	0.02	0.16	0.04	0.32
Public administrations	0.06	0.04	0.13	0.06	0.29
Hospitals	0.13	0.11	0.46	0.05	0.75
Sum (weighted)	0.10	0.05	0.11	0.02	0.28

[a] Split for lawyers, architects, small private health services, private agencies, etc.

OTHER take on the value of 1, respectively. Otherwise they are 0. The percentage shares of these variables across the sectors for the observations included in this study are displayed in Table 7.2. Thus, the expected sign of the parameter estimates associated with these variables is negative. The last column in Table 7.2 lists the percentage of organisations which carried out an energy audit for each sector. Since audits may, in general, help overcome barriers to energy efficiency, the expected signs of the parameter estimates associated with the "AUDIT dummies" are negative.

The last column in Table 7.2 indicates that about 28% of the organisations in the sample had had an energy audit conducted, but the share varies significantly between 12% in the services sector and the interior construction sector and 75% for hospitals. The majority of audits were carried out by engineering firms and utilities. The quasi-public sectors (hospitals, public administrations, and non-commercial organisations) tend to prefer engineering companies, while companies from the manufacturing sectors in the sample which are usually smaller, tend to rely on utilities. Comparing Table 7.1 with Table 7.2 suggests that there is a positive correlation between the energy-intensity of the sectors and the share of organisations that had an energy audit conducted.

7.3.4 Sub-Sector Dummies

Since the commercial and services sectors are quite heterogeneous a dummy variable is included for each sub-sector. To prevent singularity of the regressor matrix, a constant is not included.

7.4 Results

To empirically assess the relevance of the various determinants for energy-efficiency barriers, separate Logit and Probit models are estimated for each barrier. The estimation results for the six equations appear in Table 7.3 for the Logit model and in Table 7.4 for the Probit model.[4]

The comparison of Table 7.3 and Table 7.4 shows that the results for the Logit and the Probit models are very similar. The parameter estimates which are statistically significant for one model are also statistically significant for the other model.[5] Since the logistic distribution and the normal distribution differ at the tails, and since the sample is rather large, the observed differences for the parameter estimates are to be expected (Maddalla, 1983, p. 23). Multiplying the slope parameter estimates from the Logit model by $3^{1/2}/\pi$, or – as suggested by Amemiya (1981) – by 0.625, yields values that are very close to the parameter estimates obtained from the Probit model. In short, the estimation results appear to be fairly robust.

In general, all estimation results are consistent with the hypotheses developed earlier. In particular, all significant parameter estimates for ENERGY, SIZE, and the AUDIT dummies exhibit the expected negative sign. First, the findings for

Table 7.3 Logit estimation results on barriers to energy efficiency

	TIME	EKNOW	INFO	PRIORITY	UNCERT	RENTED
ENERGY	–0.079*	–0.102*	–0.015	–0.028	–0.000	–0.191**
	(0.058)	(0.055)	(0.059)	(0.055)	(0.055)	(0.067)
SIZE	–0.002*	–0.001*	–0.001*	0.000	–0.001*	–0.013**
	(0.001)	(0.001)	(0.001)	(0.000)	(0.000)	(0.003)
UTILITY	0.502**	–0.444*	–0.100	–0.193	0.152	–0.607*
	(0.176)	(0.174)	(0.185)	(0.167)	(0.168)	(0.239)
ASSOC	0.854**	–0.745**	–0.552*	–0.210	–0.041	0.071
	(0.275)	(0.272)	(0.297)	(0.237)	(0.234)	(0.330)
ENGIN	0.621**	–0.572**	–0.528*	–0.064	–0.304*	–0.796**
	(0.185)	(0.179)	(0.209)	(0.171)	(0.168)	(0.283)
OTHER	–0.731*	–0.099	–0.400	–0.571*	–0.532*	0.325
	(0.386)	(0.340)	(0.409)	(0.330)	(0.323)	(0.427)
–2 log likelihood	2,400	2,427	2,200	2,433	2,454	1,865
N	1,822	1,814	1,822	1,822	1,822	1,822
Pseudo-R^2	0.05	0.04	0.03	0.04	0.02	0.10

Standard errors are given in parentheses

*Individually statistically significant at least at 10% level

**Individually statistically significant at least at 1% level

[4] To save space, the parameter estimates for the sub-sector dummies do not appear in Tables 7.3 and 7.4. They are available from the author upon request.

[5] "Statistically significant" as used in this Section means significant at least at the 10% level.

Table 7.4 Probit estimation results on barriers to energy efficiency

	TIME	EKNOW	INFO	PRIORITY	UNCERT	RENTED
ENERGY	−0.049*	−0.062*	−0.011	−0.018	−0.002	−0.097**
	(0.034)	(0.034)	(0.035)	(0.034)	(0.034)	(0.037)
SIZE	−0.001*	−0.001*	−0.001*	0.000	−0.004*	−0.004**
	(0.000)	(0.000)	(0.000)	(0.000)	(0.002)	(0.001)
UTILITY	−0.311**	−0.272*	−0.052	−0.119	0.092	−0.377**
	(0.107)	(0.106)	(0.110)	(0.104)	(0.104)	(0.134)
ASSOC	−0.499**	−0.448**	−0.315*	−0.127	−0.026	0.049
	(0.160)	(0.160)	(0.169)	(0.146)	(0.146)	(0.184)
ENGIN	−0.375**	−0.349**	−0.307*	−0.041	−0.191*	−0453**
	(0.111)	(0.109)	(0.120)	(0.106)	(0.104)	(0.250)
OTHER	−0.436*	−0.036	−0.241	−0.355*	−0.334*	0.225
	(0.224)	(0.205)	(0.234)	(0.203)	(0.202)	(0.245)
−2 log likelihood	2,402	2,427	2,200	2,433	2,454	1,876
N	1,822	1,814	1,822	1,822	1,822	1,822
Pseudo-R^2	0.05	0.03	0.03	0.04	0.02	0.10

Standard errors are given in parentheses

*Individually statistically significant at least at 10% level

**Individually statistically significant at least at 1% level

ENERGY and SIZE are discussed in more detail for the various barriers. As expected, since they have a stronger economic incentive to actively search for measures to save energy costs, lack of TIME appears to be less of a barrier for more ENERGY-intensive organisations. Also, for larger organisations, which tend to have not only more personnel but also more specialised personnel for energy management, lack of TIME does not seem to be a barrier. As for information about the split of final energy consumption, organisational SIZE appears to matter, whereas ENERGY consumption does not, since it is not statistically significant.

It comes at somewhat of a surprise that there appears to be no difference between ENERGY-intensive and less energy-intensive organisations when it comes to INFOrmation about energy saving measures. But the findings suggest that lack of information about these measures is a barrier in smaller organisations. In that sense, the findings rationalise information programmes about measures to improve energy efficiency that target smaller companies. Neither the parameter estimates associated with ENERGY nor those associated with SIZE turn out to be statistically significant for PRIORITY. Hence, when controlling for the other variables included in the regression equation, these findings do not support the hypothesis that smaller organisations' decision-making or their limited access to the capital market are biased against investments in energy efficiency compared to larger organisations.

According to the estimations, there is no statistically significant difference between ENERGY-intensive and less energy-intensive organisations' consumption when it comes to UNCERTainty about future energy prices being a barrier to energy efficiency. By contrast, the findings for SIZE suggest that uncertainty is a relevant barrier for smaller organisations. Larger organisations are likely to have more in-house expertise for managing uncertainty and risk than smaller organisations. Finally, split incentives, as captured by the variable RENTED, appear to be less of a problem in larger organisations, or in organisations with higher energy consumption. Since investments in energy efficiency pay off faster when energy costs are high, tenants are more likely to recover the energy cost savings. For larger organisations, the landlord/tenant problem may be less relevant, because they tend to own their buildings. Similarly, leases may run longer for larger organisations than for smaller ones, which facilitates the appropriation of cost savings from investments in energy efficiency.

The findings imply that carrying out an energy audit reduces all the barriers considered in this section.[6] A closer look at the effectiveness of organisations that carry out the audits reveals that audits by engineering firms reduce all the barriers analysed, apart from PRIORITY. By contrast, audits conducted by utilities are somewhat less effective since they do not appear to reduce lack of information about energy measures or uncertainty. A likely reason is that utilities' consulting efforts tend to focus on the tariff structure rather than on technical measures. Audits by industry sector associations may be less effective than engineering firms, because they tend to provide general information rather than information specific to the organisation.

7.5 Conclusions

The empirical findings presented in this chapter support the view that energy audits help to reduce barriers to the diffusion of energy efficient technologies and measures in organisations where energy costs are low. For the German commercial and services sectors, energy audits help reduce barriers such as lack of time, lack of information about energy consumption patterns, lack of information about energy savings measures, organisational priority setting, uncertainty about energy costs and split incentives. Thus, the results rationalise support programmes for energy consultations as effective instruments to accelerate the diffusion of energy efficient technologies. Interestingly, the findings also suggest that not all external energy efficiency consultants are equally effective. Engineering firms appear to be more successful than industrial sector associations or utilities. The most likely explanation is that information provided by industrial sector associations may be too general, and that utilities tend to focus on tariffs, rather than on technological or organisational measures to save energy costs.

[6] Likelihood ratio tests were carried out to test the Null hypothesis that all four parameters associated with the four AUDIT-dummies are zero. In all six cases, the null hypothesis could be rejected at the 10% level, for TIME, EKNOW, and for INFO and RENTED even at the 1% level.

Chapter 8
Innovation Effects of Regulation – Case Study for Wind Energy

8.1 Introduction and Methodology

The case of wind energy has received increased interest in the literature lately. It is seen as a promising case for radical technical change, in which a traditional technological trajectory is substituted by a new technological trajectory even under the conditions of high path dependency. However, wind energy also makes a particularly good example for analysing the interaction of regulation, innovation and their impact on competitiveness in global markets:

- The traditional aspects of regulation with regard to typical problems of innovations, such as standardisation, intellectual property regimes, or (external) spillover effects of R&D as justification for technology policies, also apply to wind energy.
- Within electricity supply, there are also various aspects of externalities, which call for environmental and safety regulations. Thus, innovations in these fields face a second externality problem leading to a double regulatory challenge. The demand for new technologies and the pressure to innovate are much more driven by regulatory action than in other fields.
- Some of the key actors involved in wind energy are operating under very specific market conditions, which became prominent under the heading of natural monopolies or more precisely as monopolistic bottlenecks. Even after privatisation and liberalisation of electricity markets, these actors are subject to specific economic regulation in one form or another (e.g. regulation of access to the grid, control with regard to monopolistic behavior).

To sum up, sustainable innovations in infrastructure fields with monopolistic bottlenecks face even a triple regulatory challenge. This triple regulatory problem makes the case of wind energy a very interesting example to study the interaction between regulation and innovation.

In Chap. 5, different methodological approaches to study the interaction of policies and innovation have been described. The system of innovation approach is the state of the art for case studies in innovation research. However, it so far has been not reflecting the specific needs of the triple regulatory challenge. Indeed, the results

R. Walz and J. Schleich. *The Economics of Climate Change Policies*
Sustainability and Innovation,

from studies which are based on natural monopoly and environmental regulation indicate that it is necessary to look into the details with regard to the electricity specific forms of implementation of the regulation or the specific role of the various energy policy instruments. However, these studies typically lack the broader concept of an innovation system, and more often are based on an outdated linear model of innovation. Thus, it seems necessary to link the research tradition in natural monopoly and environmental regulation more explicitly with the systems of innovation tradition.

This case study starts from a systems of innovation approach and distinguishes different innovation function.[1] In so far it follows the approach of Bergek and Jacobsson (2003). In addition, however, it aims at using the functions of an innovation system as a bridge to incorporate the various paradigms of the effects of environmental and natural monopoly regulation explicitly (see Fig. 8.1). It claims that in order to fill the broad framework of a system of innovation approach, it is necessary to use a heterodox approach which draws on the different paradigms used to analyse regulation so far. It is the rationale of this heterodox approach, that the paradigms from economics and policy research can be used to explain the effect of environmental and natural monopoly regulation on the functions of the innovation system, leading to a better understanding of the complex interplay between innovation and regulation.

The analysis is performed in various steps. The first step consists in the definition of system, which is performed in the latter part of this section. In the second step of

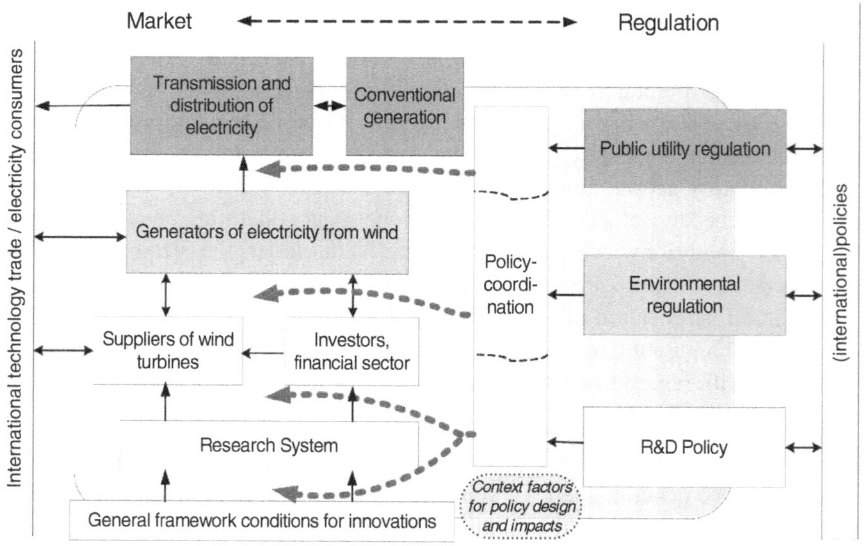

Fig. 8.1 The triple role of regulation within the innovation system of wind energy

[1] This chapter is a revised and updated version of Walz (2007).

analysis, the regulation is described. This working step gives the background for the application of the paradigms in the field of environmental and sector specific economic regulation. The output of the innovation system is measured in the third step of analysis. There are many forms how the output of an innovation system can be measured.[2] In this case study, two major indicators are used: the technical development of wind turbines with regard to turbine size and average costs one the one hand, and the diffusion of wind turbines measured in MW installed on the other. Finally, the interaction between regulation and innovation is analysed. It is in this section, there the functions of an innovation system are used as bridge to incorporate the explanations offered by the paradigms presented in Chap. 5 into the analysis.

According to Carlsson et al. (2002), the definition of an innovation system, which involves decisions on the level of analysis and the identification of the relevant actors, is always somewhat arbitrary, and necessitates various expert judgments. The following judgments were made:

- The system boundaries are defined by the development and use of a wind energy within a country. Actors with regard to both the knowledge base and with regard to diffusion of technology were taken into account.
- The analysis is performed on a national level for a certain technology, even though it has to be acknowledged that foreign markets are important for the development of wind energy, too. Nevertheless, in general, the national level is still seen as an appropriate level of analysis, because many conditions systematically differ between countries.[3] This point has to be underlined for the case at hand, because the national level still is very helpful in defining different regulatory regimes. However, in cases where there are important regional differences, e.g. in the regulatory approach between states in the US, a disaggregated look will be performed.
- In the case of wind energy, the differences between a technological and sectoral innovation system become blurred: It is somewhat arbitrary to decide whether this it is a case of a sectoral innovation system, because the output of the electricity sector is affected, or whether it constitutes a case for a technological systems of innovation, because wind turbines are build and operated also outside the traditional electricity sector.

Figure 8.1 shows the results of delineation of the most important actors for the case of wind energy: First, there are the suppliers of wind turbines. They consist of companies which have a quite similar structure such as other companies within the investment good sectors. Secondly, there are the investors in wind power. They consist of the owners of the site, together with the capital owners, which typically are private investors, wind energy funds, and in some instances electric utilities. Thirdly, the electricity produced by the wind turbines must be transmitted and distributed to the customers. Thus, access to the grid is vital for wind power. Here the electric utilities play a key role. They are responsible for the transmission and the

[2] See Grupp (1998) or Carlsson et al. (2002).

[3] Edquist (2005).

distribution of electricity on the one hand, but on the other, electricity from wind is substituting electricity supplied from other conventional power plants. Thus, the electric utilities are at the same time a competitor of wind power. Figure 8.1 also highlights the prominent role of the triple regulatory challenge in the system. Besides the direct influences on the actors affected, there are also indirect effects, as the direct influences are transmitted via the interactions of the actors with each other.

8.2 Regulation in the Wind Energy Sector

8.2.1 Overview of Instruments

There are different regulatory instruments (Table 8.1) used in most European countries and the US to foster the development of wind power. The most important instruments are:

- support for R&D,
- direct subsidisation of the installation of wind power, e.g. by tax measures,
- fixed feed-in tariffs, and
- quotas/bidding systems.

These instruments can be attributed to the different forms of regulatory challenge. R&D programmes aim at spurring innovation in wind turbines. They clearly fall into the category of regulation of R&D. Tax measures aim at improving the economy of wind energy. They are a classical "second-best solution" used in environmental regulation if the first best solution – the internalisation of "all" external costs – is politically infeasible. Fixed feed in-tariffs and bidding schemes/quotas tackle two issues: first, they ensure that the price paid to the producers of electricity from wind energy is above the price for conventional electricity. Secondly, they also aim at the integration of wind energy into the electricity grid and facilitate access to the monopolistic bottleneck. Thus, they are a key instrument which falls under both categories, economic and environmental regulation. Indeed they are used in most European countries and the US in one form or another.

In addition to the regulatory instruments aiming specifically at wind energy (or at least production of electricity from renewable energies), there are also regulatory measures which are important for the electricity sector as such. Here, many countries such as the EU or the US have experienced the trend towards liberalisation of the market. Furthermore, there are various other energy policies aiming at a rational use and conversion of energy, e.g. energy taxes and, most recently, emissions trading in the EU. However, due to political restrictions, these policies aim at changing the relative prices between technological alternatives with small cost differentials, but not with a substantial internalisation of the external costs. Thus, their impact on creating a level playing field for wind energy has been rather low.

Table 8.1 Instruments used in EU countries to foster wind energy

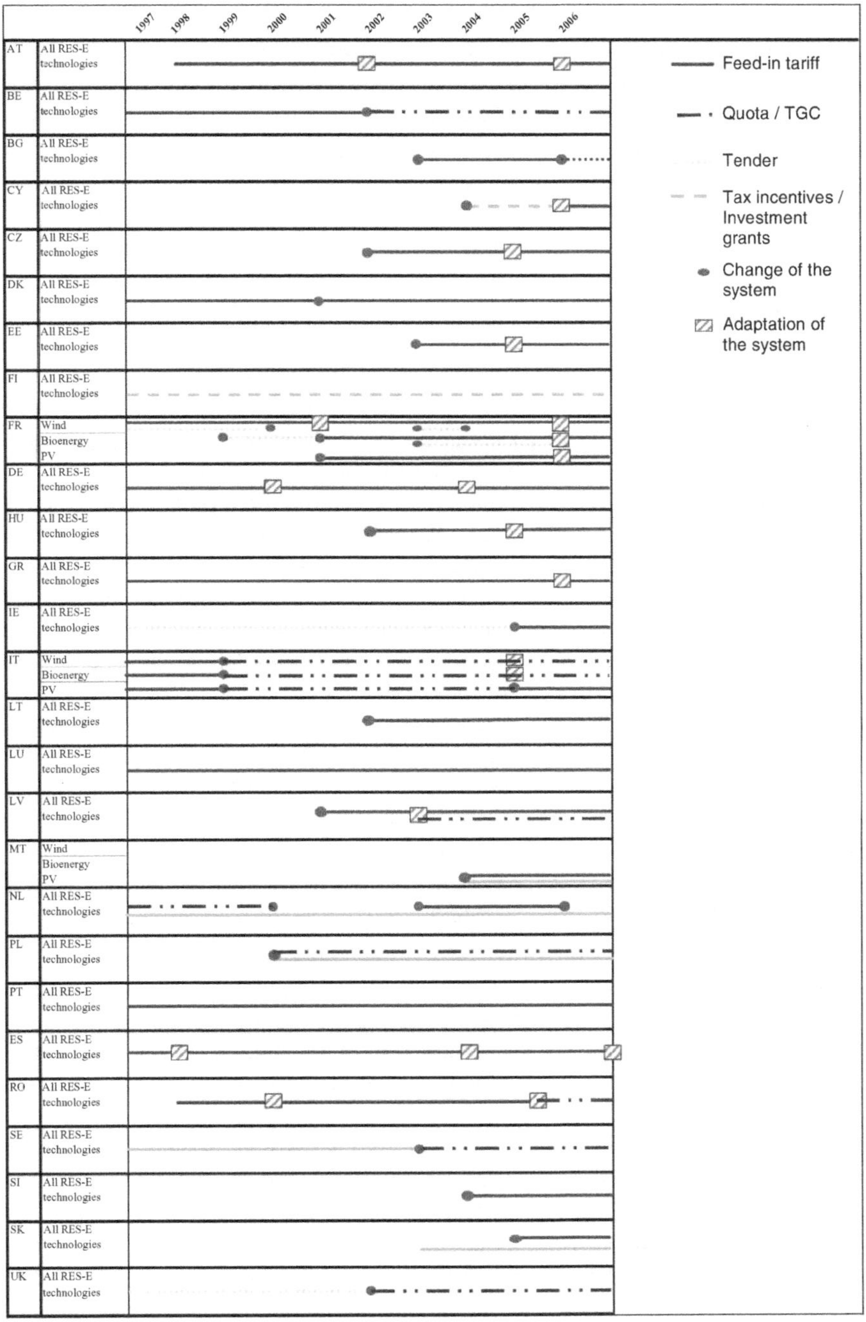

Source: Ragwitz et al., 2007

8.2.2 Regulation in Germany

In the 1980s, there was a substantial R&D programme of the Federal Department for Research and Technology (BMFT), which supported different types of wind turbines (horizontal and vertical axis, different number of blades), increasing the variety of options. Furthermore, the research programme subsidised investment in wind turbines through various demonstration projects. In 1989, a market stimulation programme was introduced, which called for an installation of 250 MW of wind power. It guaranteed a fixed payment per kWh of electricity produced, together with investment subsidies for private operators such as farmers. This programme was effective until 1995.

Based on the insights from regulatory economics, the Electricity Feed-in Act was introduced in 1991. It mandated that grid operators paid 90% of (average historical) electricity retail prices as feed-in tariffs for electricity generated by certain Renewable Energy Sources (RES) such as wind. Furthermore, it required utilities to accept the electricity delivered by wind turbines.

Due to the success in renewable energy diffusion, the Electricity Feed-in Act in its later stage had a cap to prevent very uneven burdens for regional grid operators: a grid operator had to pay these feed-in prices until the share of electricity from RES reached the cap of 5%. Nevertheless, this regulation still affected the utilities operating the grid asymmetrically. Wind turbines which benefited most under the Energy Feed-in Law are concentrated in Northern Germany. Thus, grid operators in the North felt at a competitive disadvantage, which created a problem, especially once electricity market liberalisation began. Furthermore, the falling electricity (retail) prices resulting from liberalisation also led to lower feed-in prices for electricity from RES. This started to undermine the economic base of the numerous wind turbines which had been installed in the previous years. Thus, an intensive debate emerged about the future of the Electricity Feed-in Act.

This debate was influenced by the political goals of the EU and the German government with regard to renewals. The German government had stated its intention to increase the share of renewals in electricity supply to 12.5% in 2010, and to at least 20% in 2020. Clearly, wind is seen as one of the cornerstones necessary to reach these goals. This strengthened the legitimacy of the technology, and the need for regulations to reach that goal was reinforced. In the end, the Renewable Energy Act (REA) of Spring 2000 replaced the Electricity Feed-in Act. In 2004, the original REA was amended.[4]

As a consequence of the developments described above, under the REA, feed-in prices are no longer linked to electricity retail prices, but fixed for 20 years. The cap on the share of electricity from RES was abolished. As a consequence of the ongoing diffusion of wind power, the total amount of feed-in subsidies will be distributed evenly among all high voltage grid operators, leading to levelling out of

[4] BMU (2004).

the REA payments between all of the utilities. This mechanism of passing on the subsidised feed-in prices to consumers is rather complicated. The mechanism works on three levels: first, the utilities buying RES pass on the feed-in payments to the level of the transmission network. Secondly, the operators of the transmission network average out the amount of RES and feed-in payments among themselves. As a result, every transmission operator bears the same share of RES and feed-in payments. Thirdly, the operators of the transmission network pass on the costs to the utilities performing distribution or to the electricity traders according to the electricity shares delivered. As a result, every utility or electricity trader serving a final customer receives the same percentage of RES (the so-called REA-quota) at a uniform averaged feed-in price for Germany. Thus, this levelling out mechanism ensures there are no detrimental effects on the competition between utilities.

The REA is in principle a subsidy with respect to the favoured group (the RES producers), but with the special feature of financing by the end-users of electricity. This leads to an increase in the average price of electricity, which is currently estimated at about 0.3 €-Ct/kWh for all forms of RES. The past and probable future success in diffusion has also led to changes in the regulation: in order to curb the effects of increasing electricity costs on the electricity-intensive industries, certain industries were exempted from the levelling mechanisms in 2003. With the amendment of the REA in 2004, the number of exempted industries was increased. Under the regulation, companies with an electricity consumption above 10 GWh/a and an electricity cost share above 15% in gross value added are eligible for the exemption.[5]

In the REA, a twofold degression of the feed-in tariffs is implemented; furthermore, there is a differentiation according to the site characteristics:

- In order to retain the incentive for technology producers to offer more efficient products every year, there is a reduction of the tariffs from 1 year to the next. In the 2004 amendment, this reduction of tariffs was set at 2%.
- The feed-in tariffs are substantially lowered (by about one third) after the initial period of installation.
- There is a differentiation between different sites for wind power. The switch to the lower rates usually takes place after 5 years. However, off-shore sites receive the higher initial feed-in prices for up to 12 years. Furthermore, they receive higher feed-in prices if they are constructed before 2010. On the other hand, sites with below average wind yield (less than 60% of the reference value) are not eligible for the REA at all.

The feed-in tariffs are reviewed every 2 years in the light of technological and price developments and feed-in tariffs for newly installed sites can be changed accordingly (Table 8.2). For every installation, the date of *expiration* is 20 years after the date of installation.

---

[5]BMU (2004).

Table 8.2 Feed-in tariffs for newly installed wind turbines in the year indicated

Year of installation	Calculated feed-in tariff according to old Feed-in Act	First years after installation (5 years or longer)[a]	Following years
1991–1999	90% of average price (8–10 €cent/kWh)	–	–
2000–2001	–	9.1	6.2
2002	–	8.96	6.11
2003	–	8.83	6.02
2005	–	8.53	5.93
2010	–	7.71	4.87
2012	–	7.40	4.68
Annual change in feed-in tariff	–	–2.0%	–2.0%

[a]For wind power plants exceeding 150% of yield of reference plant

The fixed feed-in tariffs of the EEG are set above the avoided costs. However, it is difficult to calculate the difference exactly. First of all, due to the reduced rates in the second phase, the average feed-in tariff for a wind power plant depends on the plant's lifetime. Second, it is difficult to calculate the avoided costs of using transmission and distribution lines which are likely to occur. Nevertheless, a first estimation of the difference between avoided costs and feed-in tariffs is in the order of magnitude of perhaps 3–4 € cents/kWh. However, it has to be taken into account that wind energy is associated with much lower external environmental costs than is conventional electricity supply. Indeed, it can be argued that the feed-in tariffs are in the order of magnitude of the avoided costs if estimations of the external environmental costs are taken into account.[6] Thus, according to environmental economics, the REA could be justified in moving renewable energies towards a more level playing field.

8.2.3 Regulation in the US

There is very little in the way of federal regulation of the wind power industry in the United States. Most innovation comes from financial subsidies, incentives, and research partnerships. As wind power is virtually non-polluting, there has been little regulation of it. Most regulations from the EPA stem from environmental and health risks, and as there are no true risks associated with wind power, there has been no need for regulation of this kind.

[6] Hohmeyer (2002).

Support for R&D is an important form of regulation in the US with regard to wind energy. The US Department of Energy's (DOE's) Office of Energy Efficiency and Renewable Energy manages the federal wind energy programme in accordance with national energy policy. Wind energy diversifies the nation's energy supply, takes advantage of a domestic resource, and is seen as a technology which helps to curb emissions of greenhouse gases. The Wind Energy Programme supports this mission by working with members of the wind industry to research and develop advanced, low wind speed turbines that will reduce the cost of wind energy in broader regions of the United States. Furthermore, research is conducted at the National Renewable Energy Laboratory (NREL) and at the National Wind Technology Center (NWTC).

In addition to technology policies, there have been several policies affecting wind energy in the past in the areas of regulation of the monopolistic bottleneck and environmental regulation. The Public Utility Policy Act from 1976 stated that the feed-in tariffs from power stations outside the electric utility industry had to be charged at avoided costs. Among others, this also applied to wind energy. Especially in times with rising marginal costs, this provoked an intensive debate about whether or not feed-in tariffs above average costs were justified.[7] Tax incentives for renewable energy were another important policy which helped to promote the boom in wind energy in California in the late 1970s and early 1980s. Furthermore, the Department of Energy (DOE) spent substantial amounts of money for R&D projects. However, all these incentives were drastically reduced under the Reagan Administration in the 1980s.[8]

A key federal policy to foster wind energy is the Federal Energy Policy Act of 1992. The federal government provides a production tax credit of 1.5 cents per kWh (adjusted for inflation this is equivalent to 1.8 cents in today's terms) for electricity generated by a wind plant during its first 10 years of operation. This credit is intended to "level the playing field" for wind, which must compete with other energy industries that receive billions of dollars in federal subsidies each year. In addition to the federal level, there are different state programmes such as tax incentives (e.g. exemption from state or local sales taxes or property taxes), direct cash incentives, and low-cost capital programmes (subsidised loans and loan guarantees). The federal production tax credit expired at the end of 2003 and was only extended in October 2004 for an additional year. Thus, there has been substantial uncertainty about the future policy, and the US wind energy industry has been calling for a long-term extension so that companies can plan under steadier framework conditions in the future. In 2005, the scheme was renewed for 3 years. Thus, the uncertainty has been reduced temporarily.

On the federal level, there has been considerable discussion on the implementation of a nationwide Renewables Portfolio Standard (RPS). RPS is a "minimum content requirement", which specifies that a certain minimum percentage of electric

---

[7] Walz (1995b).
[8] Brauch (1996).

power must be generated from renewable energy sources (wind, solar, and others). Typically, RPS legislation provides that the minimum percentage increases gradually over time to encourage the sustained, orderly development of the renewable energy industries.

Renewable Energy Credits ("RECs") are central to the RPS. A REC is a tradable certificate of proof that one kWh of electricity has been generated by a renewable-fuelled source and sold to an end-user. The RPS boils down to a requirement that every generator should possess a number of RECs equivalent to a determined percentage of its total annual kWh generation (or sales). For example, if the RPS is set at 5%, and a generator sells 100,000 kWh in a given year, then it would need to possess 5,000 RECs at the end of that year.

For generators that fall short of the required number of credits at the end of the reporting period, an automatic penalty for non-compliance is calculated. The penalty is three times what it would have cost to purchase each REC that the generator failed to acquire. This penalty is estimated to be about 3¢–5¢ per REC, high enough to encourage full compliance, yet not so high as to encourage litigation. The high penalty level is intended to make the policy self-enforcing by avoiding having to resort to costly administrative and enforcement measures, which would induce additional transaction costs.

The US Senate approved the Renewables Portfolio Standard (RPS) as part of the Energy Bill S. 217 in 2002. However, the Energy Bill which passed the House of Representatives (H. R. 4) did not include a similar RPS. A new Energy Bill was introduced during the 108th session of Congress.

Even before national legislation has become effective, some states have already autonomously introduced renewable portfolio standards. Currently there are 25 states with such a scheme, among them California and Texas. The stringency of the standards and the time horizon differs greatly between the states. The RPS of California has received the most attention. This mandates that a 20% renewable standard is reached. Other states have mandated similar targets, with Oregon and Minnesota even aiming at 25%, however not before 2025.

8.3 Development and Diffusion of Wind Turbines

8.3.1 Development of Technology

The design of wind turbines can be traced back for decades, but it was not until the dramatic oil price increase during the 1970s and early 1980s that the development of wind turbines was incorporated into the energy policy agenda. However, the situation in the 1980s was dramatically different to the one observed today. In the 1980s, no dominant design had been developed. There were a variety of turbines being experimented with, using both horizontal and vertical axes. At the same time, the number of blades ranged from one to more than three. There were a number of

Fig. 8.2 Milestones in wind turbine development (Source: AWEA, 2005, p. 15)

small players experimenting with small-scale wind turbines. Some larger companies also entered the field and attempted to build larger turbines of MW capacity. In general, these attempts in the 1980s were not successful, exemplified by the well known 3 MW GROWIAN plant in Germany, which was erected by MAN in 1982, but dismantled again in 1987 due to its failure. Despite these drawbacks, major progress was also achieved during this period: the costs per kWh were reduced by over 50%, and a standardised wind turbine with a capacity of 55 kW emerged with the Danish Micon 55, which was installed not only in Denmark but also in great numbers in California (Fig. 8.2).

During the 1990s, the development of wind turbines continued. This was accompanied by constantly increasing turbine size. The increasing size of turbines was not only achieved in R&D demonstration projects. During the 1990s, the average size of installed turbines also increased continuously. Within Germany, for example, the average size of turbines installed from 2000 onwards is 20 times bigger than those in the late 1980s – a remarkable innovation success in little more than 10 years.

The innovations described above plus the economies of scale also had considerable effects on the economics of wind power. The average investments per kW of installed capacity decreased substantially. At the same time, the cost differences between countries narrowed. This indicates that, within the leading countries, competent suppliers have emerged which are now competing internationally, leading to an erosion of cost differences between countries. At the same time, the cost of wind power per kWh electricity produced has fallen constantly. However, the cost degression effect observed in the statistics is likely to be somewhat lower than the cost decrease in specific investments due to the fact that, with improved wind turbines, sites with lower wind yields are also used more widely.

The future perspectives of wind turbine development are characterised by further cost degressions. Current costs for medium sized turbines at a medium quality wind site are 5–6 c€/kWh. Depending on the assumptions of future growth rates, learning rates and the corresponding times for each doubling of installed capacity, the electricity costs are expected to decrease to values of 4–5.5 c€/kWh by 2010. Future challenges include, for example, the use of new materials to lower the weight of the blades, or the increased use of fuzzy controls to improve the efficiency of the turbines. Furthermore, due to the growing number of turbines operating, it is becoming increasingly important to deal with the issue of integrating them into the electricity system without affecting the stability of the grid. Another key challenge is the development of off-shore wind power. Clearly off-shore offers the perspective of a huge number of sites with very good wind yields becoming available. With access to good onshore sites becoming increasingly scarce, off-shore wind is seen as central to the future expansion of wind power. However, there are several challenges still to be met, such as the installations of turbines at sea and their anchoring on the seabed. In general, it can be argued that wind power innovations will require a greater degree of additional knowledge (e.g. material sciences, geological science) from outside the typical mechanical and electrical engineering sciences than in the past in order to be able to meet the future challenges. Thus, future developments will require a substantial broadening of the knowledge base.

The importance of future technology development can also be seen in the patenting activity. The rise in international patent has been stronger than for all patents. Furthermore, the share of the different countries at world patents also shows in which countries new knowledge mainly emerges. Here, Germany has been clearly taking the lead. It has constantly increased its shares, whereas the US has been falling behind (Fig. 8.3).

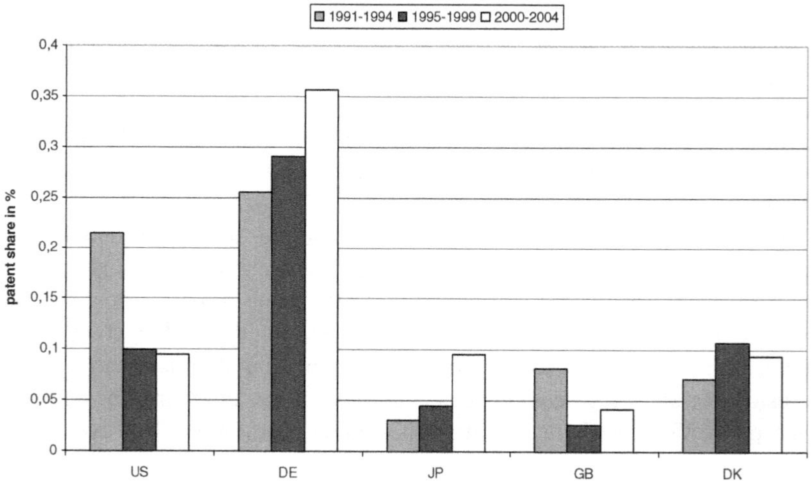

Fig. 8.3 Development of patent shares for wind power (Source: calculations of Fraunhofer ISI)

8.3.2 *Diffusion of Technology*

The emergence of the innovations described above led to a vast diffusion of wind power. The annual globally installed capacity increased continuously from less than 500 MW in the early 1990s to almost 8,000 MW in 2004 (Fig. 8.4). This resulted in a tremendous increase in the accumulated installed capacity from less than 1,000 MW in 1991 to almost 50,000 MW in 2004. This development has been mainly driven by the success of wind energy in the EU. In 2004, the EU accounted for about three quarters of world capacity, up from less than 50% in 1991. Thus, the EU has clearly taken the lead in wind power. The US, in contrast, showed only a modest increase in its installed wind power in the 1990s, with about 2,500 MW installed capacity in 2000. The development in the US only began to gain momentum at the beginning of this century, with an increase up to 8,700 MW by the end of 2005. This development was concentrated in a few states, notably California, Texas, Minnesota, Iowa and Wyoming. Recently, there has been a substantial increase in wind power development outside Europe and North America. For example, India has been increasing its accumulated wind energy installations to about 3,000 MW.

Within the EU, Germany has increased its share of the installed capacity to about 50%. With approx. 18,500 MW installed in 2005, Germany accounts for more than one third of global wind capacity. This success is all the more remarkable because it has come about in such a short period of time. In the 1980s, the diffusion of wind power in Germany was not particularly impressive. At the end of 1989, the

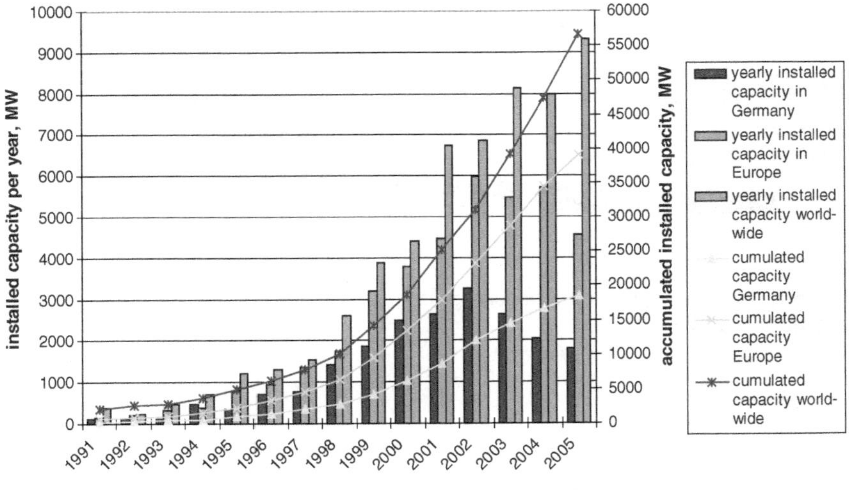

Fig. 8.4 Development of annual and accumulated installed capacity of wind turbines (Data: Ender, 2002; Durstewitz, 2003; Ender, 2004, 2005; OECD, 2007)

accumulated wind power capacity in Germany amounted to only 20 MW. Thus, Germany's pole position in wind power production has been achieved in only one decade. A similar development has been taking place in Spain, which holds the number 2 position in accumulated installed capacity in the world, and is thus even ahead of the US The role of Denmark, which championed wind energy development in the 1980s and early 1990s, has somewhat diminished in absolute terms, with about 3,000 MW installed until 2005.[9]

Comparing only installed capacities may be misleading because the electricity generated also depends on the full load hours operated at each installation. There are some differences between the European countries. For Denmark, the UK, and the Netherlands, full load hours in the order of magnitude of 2,200–2,300 h/a are common. In other countries, such as Spain, Austria and Germany, the full load hours are somewhat below this, reaching 1,700–1,800 h/a on average. Furthermore, in order to evaluate the relative success of the countries, their size differences must also be accounted for. Looking at the percentage of installed wind capacity in total capacity, Denmark clearly has the highest percentage of wind power, followed by Spain and Germany with each about 7%. With regard to this indicator, the US lags far behind, with about 1% of its capacity coming from wind energy (Fig. 8.5).

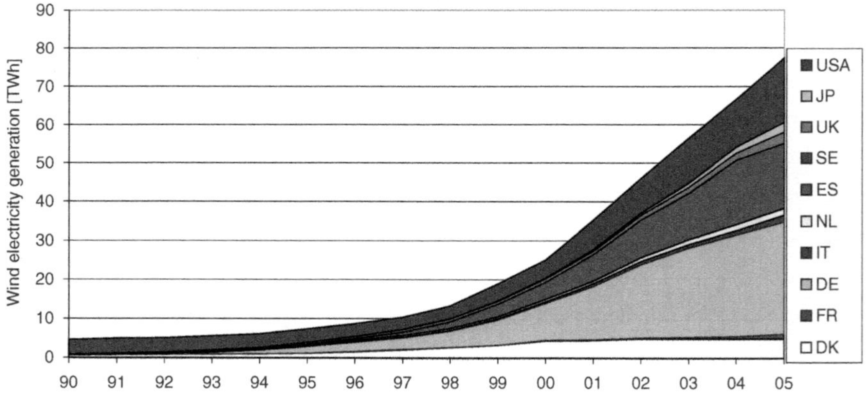

Fig. 8.5 Deployment of wind electricity generation in selected countries 1990–2005 (Source: Ragwitz et al., 2007)

[9] For the special case of Denmark over time, see Smith (2003) and Meyer (2004).

8.4 Comparative Analysis of the Influence of Regulation on Innovation

8.4.1 Experimentation Phase in the 1980s

The comparative analysis of regulatory policies reveals both differences and similarities between the US, the EU and Germany in particular. In all countries, R&D policies are used widely as an instrument to spur knowledge creation in the early phase of experimentation. However, there are substantial differences with regard to the interplay with other functions of the innovation system.

In the US, the federal government's involvement in wind energy research and development began in earnest within 2 years after the so-called "Arab Oil Crisis" of 1973. Furthermore, the PURPA legislation established, in principle, access to the grid. Indeed the US showed first signs of success with the diffusion of wind power which was symbolised by the early diffusion of wind farms in California.

Despite these early moves, the different forms of regulation ultimately proved to be largely ineffective because of the interference of political factors and the withdrawal of financial support before success was achieved. The US scaled down its efforts by about 90% in the 1980s, marking a substantial break in the continuity of the policy. As a result, the legitimacy of wind energy was undermined. There were no political goals or debates strong enough to counteract the effect of falling oil prices, which led short-term development away from alternative energy supplies. Furthermore, a substantial part of the Californian wind energy boom was supplied by Danish firms. Within the US, no network of suppliers had been developed, and the basis for variety was rather small. Thus, at the beginning of the 1990s, there was no national wind supply industry in the US powerful enough to participate in the subsequent world-wide boom.

During the 1980s, wind power in Germany was in the experimentation phase. Knowledge creation and guiding the direction of search was substantially facilitated by the German R&D programme.[10] A wide range of new technical knowledge was provided because the R&D programme was not limited to one design only. There were many new entrants experimenting with wind turbines. The early legitimacy of wind power was secured by a comparatively strong environmental movement. The debate on an alternative future for the electricity industry was continued even after energy prices fell in the aftermath of the second oil price crisis. This can be attributed to an intensified debate about nuclear phase-out after the Chernobyl accident and the start of the global warming debate at the end of the 1980s.

[10] Hemmelskamp (1998), Lauber and Mez (2004).

Furthermore, the expansion of neighbouring Danish firms in the 1980s who exported numerous wind turbines to California gave another signal for market entry. Finally, a network of suppliers with user linkages emerged around the R&D programme, enabling the exchange of information and knowledge and positive economic externalities. Thus, the experimentation phase, supported by the R&D policy, provided the German wind power innovation system with variety, numerous players and accumulated knowledge and competence to be built upon during the subsequent phase of market expansion.

8.4.2 Rapid Market Growth in the 1990s

These different supply side starting points at the beginning of the 1990s were supplemented by different policies to foster diffusion: In the US, the primary policy was a subsidy in the form of a tax credit, introduced in 1992. In some European countries including Germany, the prime instrument was a fixed feed-in tariff above the avoided costs of utilities. It is difficult to compare the intensity of subsidisation between the countries. The difference between feed-in tariffs and avoided costs in Germany seems to be larger than the tax credit per kWh granted in the United States. However, for a more comprehensive comparison, the effects of the Renewable Portfolio Standards implemented so far and the state subsidies would have to be included on the US side. On the other hand, the effects of the tax subsidies on investors in Germany have to be accounted for as well.

A very important difference is the payback predictability in schemes using fixed feed-in-tariffs (e.g. Germany and Spain in Europe) and countries using bidding systems/quotas (e.g. the United States and a few European countries such as the UK, Ireland and Italy). The analysis for Germany and the US revealed that payback predictability was identified as a key issue for the availability and cost of capital to the investors. Especially small-scale investors, who have to refinance their funds through the financial system, claim that the predictability of the feed-in tariffs is essential for securing the funds they need. The need for a differentiated treatment of small-scale investors and large-scale utilities can be explained by different transaction costs and problems in changing routines within the financial institutions when faced with the new opportunities of independent power production. In contrast to the situation in the US, investors under a fixed feed-in-tariff scheme are able to present predictable paybacks for their investments to the financial institutions. Thus, fixed feed-in tariffs reduce the risk of fluctuating prices for the investors. The analysis revealed that German investors are much more likely to receive private funds at normal capital costs compared to the investors in the US, who are much more likely to be paying premium capital costs.

This also helps to explain some of the differences in wind power development within the EU. Clearly fixed feed-in tariffs fulfil the function of supplying financial resources much better than quota or bidding systems. Thus, it is not surprising that the countries leading wind energy diffusion (e.g. Germany, Denmark, Spain) have been relying on that kind of regulation. Moreover, the weak diffusion of wind power in the UK, despite its superior sites, can be attributed to the short term security offered by the UK policies to investors.[11] The difference between the two schemes might become even more important if the volatility of electricity prices increases with the liberalisation of the electricity markets.

In general, it is argued that the assurance of fixed feed-in tariffs exerts less pressure for future innovation than do green certificates or bidding systems.[12] The argument is that only the lowest cost solutions are able to receive the benefits from these systems. However, Haas et al. (2004) and Meyer (2004) argue that the situation is more complex, with all of these instruments being a mixture of market pressure and regulation. For example, they offer evidence that the prices under the UK renewable obligation scheme have reached the level of the German fixed feed-in tariffs, without having been able to improve diffusion as much. Indeed, it can be argued that a carefully designed fixed feed-in scheme, which assigns lower feed-in prices to resources or sites with lower generation costs, might be able to reduce the producer rent of the suppliers of renewable energy in favour of a lower price to the customers. Furthermore, tailoring fixed feed-in tariffs to the needs of the different resources and technologies opens up the possibility of bringing a greater variety of technologies into the market. Thus, the diversity of technological solutions may increase compared to a quota system.

The specific form of the German Renewable Energy Act also provides incentives for dynamic economic efficiency. There is a constant incentive for wind turbine producers to become more efficient since, firstly, there is competition for the customers, who can increase their profits by choosing a cheaper technology, and, secondly, the feed-in tariffs decline each year for newly installed equipment. Individual installation owners also have, where technically possible, incentives for efficiency gains because this increases their profits.

In Chap. 5 it was argued that innovations do not only depend on the economic incentives of relative prices highlighted by neoclassical economics. From an innovation systems view, the communication between the various actors and the feedbacks between users and producers are important aspects, too. Innovation and diffusion are not sequential phases, but learning and future innovations depend on experiences made during market diffusion. From this point-of-view, it can be argued that the predictability of the feed-in tariffs, which led to an enormous diffusion of wind power, was also a prerequisite for the development of markets big enough to exploit economies of scale and learning curves which help to drive technology prices down towards the levels of conventional technologies of electricity production. In addition, the distinction made within the REA between different sites (onshore/off-shore) provides additional guidance in the search for new solutions and facilitates the creation of new knowledge with regard to off-shore facilities. To sum

[11] Mitchell and Connor (2004).

[12] Beise and Rennings (2005).

up, a more detailed look at the provisions of the regulation together with an innovation systems approach does not support the contention that the German REA regulation has been less effective for new technological development than a bidding system.

Differences in the supply of resources, which is a key function of an innovation system in the phase of market expansion, are, however, not the only explanation for the differences in the market growth of wind power in the 1990s. It has to be kept in mind that the different functions interact with each other, and that a take-off requires vicious circles to be established between the different functions. Taking these interaction effects into account, the following picture emerges:

In the US, the primary policy was a subsidy in the form of a tax credit, introduced in 1992. However, it took almost 10 years before substantial take-off was observed. In addition to deficiencies in fulfilling the function supply of resources, some of this delay can be explained by the lack of legitimacy of the technology, e.g. because of reluctant environmental and climate policy. Furthermore, the specific situation in the US with regard to coalition building has to be taken into account. Previous analysis (Walz, 1995b) has shown that the Public Service Commissions open up the floor for coalition building. Coalition between consumer groups fighting for low electricity rates and environmental groups has been a key element in explaining pro-environment developments such as the introduction of Least Cost Planning or the halt on building new nuclear power plants. However, the interests of consumer groups and environmentalists were clearly at odds with regard to wind energy. Thus, the process of PSC was not able to fulfil the function of advocacy alignment. It needed various additional aspects to overcome these obstacles, such as the success stories of wind energy abroad, the debate and implementation of RPS in some states (supplemented by state subsidies) and the acquisition of at least one company (GE Electric, which installs about 50% of US market) which was seen as a national champion in supplying wind turbines. Clearly, the extension of the tax credit for 3 years reduced uncertainty. Furthermore, with about half of the states having RPS, the legitimacy of the technology has been substantially improved. However, the current development is still a fragile one. The tax credit system is to expire again, and the policy proposals call for a 1 year renewal only. There are discontinuities within the policy, notably the expiration of the production tax credit, and discrepancies between the technological goals of the wind energy programme and the overriding energy policy goals which seem to favour more traditional forms of energy supply such as Arctic oil or nuclear power. Thus, there is still no coherent diffusion policy.

With the Electricity Feed-in Act of 1991, the German wind energy innovation system entered a phase of rapid market growth. The relatively high political weight attributed to environmental issues such as global warming increased the legitimacy of wind power and gave additional guidance for the direction of research. This effect is in line with the arguments of the policy analysis approach. The fixed feed-in tariffs facilitated the creation of the enormous market described in Sect. 4. The reduced risk for investors mobilised not only private capital for investments on a large scale. Part of the economic benefits spilled over to the suppliers of wind turbines which helped to supply the resources for further innovations. With available sites becoming scarce, the need for up-scaling turbines became more pronounced. Furthermore, the market expansion helped to exploit economies of scale leading to lower prices and increasing profitability. At the same time, the supply side of the wind turbine market consolidated considerably, shown by the rise in concentration levels since 1992. Among the different wind turbine producers, a few companies emerged as key leaders driving the innovations, with Enercon at the front. And even though Tacke was first bought by Enron and then subsequently General Electric, the key departments remained in Germany, and there are plans to locate the new R&D research laboratory in Germany, too. Thus, the companies Enercon, GE Wind (at least partially German based, formerly Tacke), Repower and Nordex are among the world's leading players in wind turbine production, together accounting for almost 40% of the world market in wind turbines in 2003. This market formation on the supply side reinforced the organisation of common interests and the legitimacy of the technology. This was a prerequisite for aligning the conflicting interests when the electricity prices were increased by the fixed feed-in prices. First of all, liberalisation made utilities much more wary about cost differentials among themselves, resulting in those utilities which were affected more by the scheme strongly opposing the whole instrument. This conflict was aligned by the levelling scheme of the REA. Second, electricity-intensive industries such as primary aluminium were heavily opposed to the system as soon as they felt it would threaten their competitiveness. This led to the exemption rules of 2003 and the amendment in 2004, which managed to defuse the opposition from the industries affected while keeping the system intact.

To sum up, during the phase of market growth various virtuous circles were introduced which enabled the German wind energy innovation system to better fulfil its functions, pushing generation and the diffusion of innovation further and further. Clearly, however, this phase of rapid expansion with continuous innovations would not have been possible without the market formation induced by regulation. The phase of market expansion in Germany led to virtuous circles which allowed German producers catch up with Danish ones. In sum, the German wind power innovation system had a high functionality during the take-off phase in the 1990s. The result was that, at the beginning of this millennium, companies in Germany and Spain joined the Danish ones as key players in world-wide wind turbine supply.

For various years, the demand in Germany was increasing so fast that German suppliers were mainly concentrating on their home market. Thus, they were rather reluctant to enter the world market. Indeed, the surge in demand in Germany was so strong that imports of wind turbines quickly increased. After the build up of additional capacity, and the slowing down of capacity increase in Germany, this situation changes. German suppliers have been increasingly turned towards exporting the technologies. Indeed in 2006, German exporters were closing the gap to the still leading Danish firms substantially (Fig. 8.6).

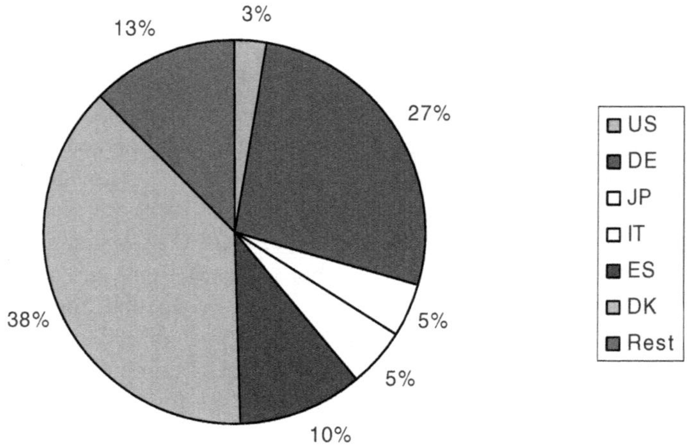

Fig. 8.6 Export shares of countries at wind turbines in 2006 (calculation of Fraunhofer ISI based on comtrade data)

8.4.3 Future Outlook

The future of wind power in the different countries is difficult to forecast. It depends, among others, on the market strategies of the actors involved and especially on the policies implemented. Within the European Union, some countries have introduced new policies. In France, for example, it can be expected that the introduction of the fixed feed-in-tariff scheme will lead to an expansion of wind power. The future of wind power in Germany will most likely be characterised by further cost reductions and expanding markets. The design of the Renewable Energy Act, with its decreasing feed-in tariffs over time, gives an additional incentive in that direction. However, it can be argued that the R&D of the suppliers of wind turbines, which at present is characterised by structures similar to engineering consultants, needs to be enhanced in order to meet the challenges described in Sect. 4. It is argued that suppliers already have problems keeping pace with technological developments. However, some of the key innovations ahead will require results from basic research being transferred as quickly as possible to application. Thus, it will most likely be necessary to increase the knowledge base of the R&D of the wind turbine suppliers by establishing much closer links to basic research institutions in fields such as material science, geology etc. Perhaps a new phase of R&D policies might be necessary, supporting the regulatory approaches used in Germany which have proved to be so successful in fostering innovation.

At the same time, the future impact of regulating the monopolistic bottlenecks might affect wind power too. So far, the German utilities were able to transfer some of their monopolistic ownership of the grid to the retail market. This can explain how even small local utilities were able to survive the first phase of liberalisation.[13] However, this situation might change with the upcoming regulation of the access to the grid by a regulatory commission. There are different possibilities how this might affect wind power:

- It can be foreseen that local utilities will be under severe pressure. One reaction might be diversification, perhaps towards the generation of electricity in regulated markets such as renewables.
- Another reaction might be that local utilities surrender to larger utilities, which would change the political economy of the power sector in favor of large utilities. The attitude of large utilities towards wind power remains unclear. They might be interested in investing in wind power themselves, e.g. in large wind farms offshore. Such a tendency could be promoted if other regulations, e.g. emissions trading rules beyond the first commitment period, support such a move (see Walz and Betz, 2003).
- On the other hand, large utilities have a growing ability to block the future development of wind power. In order to raise the share of fluctuating wind power, the stability of the system is an increasingly crucial issue. Increases beyond the current level require the cooperation of the grid operators. Furthermore, the development of large off-shore fields requires additional investments in the transmission system. Thus, the decision to expand wind power will shift away from single investment decisions for an innovative technology towards a system innovation, requiring the coordinated investments of different players.

Clearly the future development will pose new challenges for regulation. It can be foreseen that the regulation of the feed-in-tariffs, which has so far been directed mainly at small-scale investors, has to be connected with the regulatory measures of the rest of the electricity system.

For the US, one crucial point will be the design of a long-term policy strategy for wind power. If this can be achieved, there are indications that the US could assume a more important role in the world market. The takeover of the German producer Tacke by GE Wind has created an important German-American player. Perhaps most important, however, are the indications from the long-term R&D policy. If the US is successful in linking basic research and application as stated in their R&D programmes, they might be better able to tackle some of the key technological challenges lying ahead. This could substantially improve their relative technological competitiveness. If this is supported by a stable long-term policy fostering the diffusion of wind energy, the US might win a leading role in wind energy, unless European countries are heading in the same direction.

[13] Brunekreeft (2002, 2004), Walz (2002).

8.5 Questions for Further Research

The developments of the past 30 years have shown that reversals in the world leadership of wind power are possible to achieve. After an initial boom in the US in the 1970s, some European countries, notably Denmark, Germany and Spain, have now taken the lead. The analysis has shown that this can be attributed to an innovation system which successfully met the needs of the different phases of an evolving industry. Regulatory measures in the fields of R&D policies, environmental regulation and access to the monopolistic grid were responsible for this success. In addition to key functions such as market formation and supply of resources, the rapid market growth was also made possible by creating vicious cycles between the different functions of the innovation system which reinforced each other. However, it will take additional efforts to be able to continue the European success story of the last 15 years in the long run. New technological and regulatory challenges require a continuous adjustment of the relevant regulatory policies.

The analysis demonstrated that regulation is especially important in a case characterised by a triple regulation challenge in the three fields of

- spillovers of R&D,
- environmental protection, and
- access to monopolistic bottlenecks.

From a methodological point-of-view, it proved to be beneficial to use a sectoral/technological system of innovation as the heuristic framework for analysing the role of regulation. The latest development in this string of research offers the possibility to link the role of regulation to innovation by analysing how the different functions of an innovation system are influenced.

However, the drivers and feedback mechanisms which prove to be important in the interplay of regulation and innovation are foreign to the approaches of the macroeconomic models applied so far. Indeed, it will be a key challenge to construct an empirical model which is able to explain the innovations in such circumstances. Alternative modelling approaches, such as systems dynamics, which have been developed for areas other than economics, might offer a solution.[14]

Another central aspect which illustrates the international competitiveness is the importance of exports for the national suppliers of wind turbines. Denmark still exports a much higher share of its wind turbine production (Ender, 2002). Thus, it can be argued that Denmark still holds the same position of a lead market for wind power it occupied clearly in the 1980s and 1990s (Beise and Rennings, 2005). However, this argument is actually less clear than these numbers imply. Firstly, it is becoming more and more difficult to attribute a producer to one country only, e.g.

---

[14] System dynamic models have hardly been used in macroeconomics so far. One exception is the ASTRA-model (Schade, 2005), which has been developed in conjunction with sector-specific models in the transportation sector.

GE Wind incorporates the production and R&D facilities of the former German Tacke company and plans to locate its new research laboratory in Germany. Secondly, the domestic market in Denmark has not expanded much recently, leading Danish producers to concentrate on the export markets. The German market, thirdly, has experienced a boom and has been absorbing most of the capacities of the German producers until 2005. Apparently, they were not able to increase their capacity fast enough to also serve a higher share of the world market, which might have been possible given their technological competences. However, the latest figures clearly show that German producers have successfully shifted their sales towards foreign markets. They only could do so because the increasing home demand had fostered the emergence of a growth industry which was able to establish technological leads. The successful interplay of regulation and innovation was a prerequisite for the success of the German wind energy industry on the international markets. It is the task of future research to move beyond such case studies and to establish statistically based relationships between regulation, innovation and success on the world market.

Chapter 9
Summary and Conclusions

The protection of the earth's atmosphere requires substantial efforts to reduce CO_2 emissions – especially in countries with very high per capita emissions. To meet the climate target of stabilising the concentration of greenhouse gases in the atmosphere existing climate policies will have to be expanded and new policies will have to be implemented. Climate policies such as energy or emission taxes, greenhouse gas emission trading systems, subsidies, standards or information-related instruments change the incentives structure of economic agents towards more climate-friendly processes and products. They also alter the profitability of new climate-friendly technologies, leading to additional research and development efforts towards such technologies, eventually resulting in a less carbon-intensive production and consumption structure of the entire economy.

Looking first at the macroeconomic effects of climate policies, we identify three mechanisms which drive the economic outcomes: effects on costs (supply side), effects on aggregate demand, and technological effects (productivity, technological competitiveness). It can be ascertained that the total impact results from the *interaction* of the various *mechanisms* and cannot be derived from the isolated observation of individual effects. Taken together, the combination of the different effects produces a situation as characterised in Fig. 9.1: up to a certain point, a climate protection policy is likely to lead to an increase in production. However, if more than a modest reduction of CO_2 emissions is targeted, the negative effects become stronger and stronger ultimately resulting in losses in production. The effects on employment are similar. If tax policies are applied such that the revenue is used to lower the cost for labour, the positive effects on the demand for labour are stronger and the negative effects only "kick in" at a higher reduction level.

For policy making, the central aspect is how relevant the level of effects is from a political view. The following questions are of particular interest:

- At which level of CO_2 reduction do the positive effects occur; how big are they?
- Up to which level of CO_2 reduction are no severe macroeconomic losses to be expected?
- How large is the increase in employment, for example, if CO_2 reductions are achieved which do not affect production?

R. Walz and J. Schleich. *The Economics of Climate Change Policies*. Sustainability and Innovation,

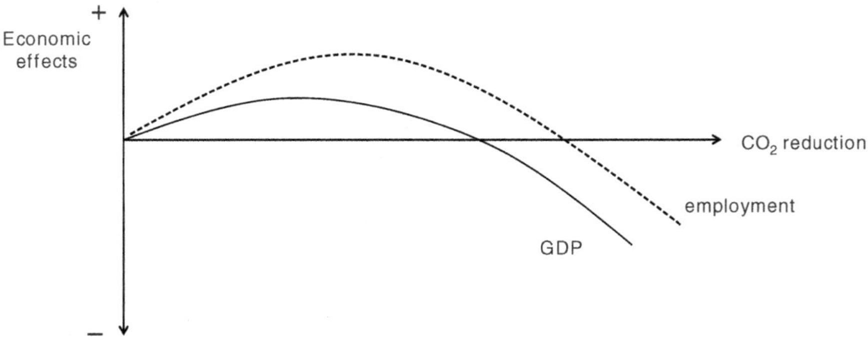

Fig. 9.1 Combined effects of the mechanisms

The theoretical analysis does not make any predictions about these kinds of question. Thus, it is necessary to turn to the empirical macroeconomics of climate change. However, different modelling approaches emphasise different segments of the numerous economic mechanisms. Thus, it can be foreseen that different results will emerge. The most important empirical studies in Germany are analysed, and differences between the results are explained. Overall, the following realistic *conclusions* can be drawn with regard to the macroeconomic impacts of climate change policies:

- A combined bottom-up/top-down approach stresses the substitution of energy by capital whereas the results derived from abstract production functions also show clear substitutions of energy by labour.
- Top-down approaches predominantly model ecological tax reform, the revenue from which is used to lower other distortionary taxes, mostly those on labour. This reinforces the employment effects in comparison to modelling a package of measures in which considerable CO_2 reductions are targeted in the space heating sector. The differences in the results can thus be explained by the fact that different energy policy strategies are being analysed.
- The extent of the spread within individual studies indicates that the actual energy policy involved is less decisive for the macroeconomic effects than how this policy is embedded within the economic policy.
- The analysis for Germany generally shows moderate macroeconomic effects. The results indicate that a short-term moderate reduction in CO_2 emissions in the order of magnitude of 10% will indeed result in an increase in employment and probably GDP, too.[1]
- In the medium to long term, a reduction of the CO_2 emissions in the order of 40% will only marginally change the GDP and bring about a moderate increase in employment. In total, a growth in the number of jobs from 200,000 to 300,000 seems to be a feasible result.

[1] This result can be perhaps characterised as symbolising the maximum point of the employment curve in Fig. 9.1.

- Combining bottom-up and top-down methods provides a good starting-point for establishing micro-macro bridges within technology-economic analyses. This approach should be refined methodologically in the future and applied more often.
- The positive effects could be amplified even more by the innovation effects, which are not sufficiently considered in the model results.

The analysis of the structural adjustments of climate change policies started with the sectoral impacts. The pattern of sectoral changes in the industry structure are analysed for two policy scenarios which differ with regard to the policy instruments involved. It can be assumed that the energy supply industry will be a loser under a climate protection policy. The differences between the studies, especially those concerning the future role of coal and natural gas, mainly reflect different assumptions about policy details. Whether or not the service sectors are among the winners also largely depends on policy details. If, for example, climate policy basically relies on a uniform CO_2 tax, the revenue of which is used to lower taxes on labour, then these sectors are likely to benefit. If the main burden of reduction is put on households and small commercial sectors and the instruments used do not lead to tax revenues lowering the costs of labour, then service sectors are not likely to benefit at all. It is clear that the construction-related sectors will be among the winners of a CO_2 reduction policy, together with certain parts of the capital goods industries.

Furthermore, the effects are analysed both on changes in the qualitative job characteristics and qualification requirements and on the regional adjustments in employment. In general, the net losses in the losing sectors tend to be more pronounced than the net gains in the winning sectors. Among the major losers are the coal-producing regions. The effects on employment in East and West Germany depend on the kind of policy applied. If climate protection policies rely on tax instruments only, West and East Germany are affected to the same degree. If a policy mix is used, however, there will be a greater positive effect in East Germany. This different pattern can be partially explained by which sectors gain or lose, e.g. the higher importance of the construction sector in East Germany. In order to measure the regional concentration, a Herfindahl Index (HI) of regional concentration has been developed. The calculations of the Herfindahl Index for the regional concentration of employment do not show much difference between the scenarios. Thus, it can be concluded that only small changes take place.

The analysis of qualification requirements reveals clear differences between a policy mix and an Ecotax. In the policy mix scenario, there is a shift from high qualification requirements towards medium requirements, with the importance of low requirements almost not affected by the policy. In the Ecotax scenario, in contrast, qualification requirements clearly increase, with a reduction in both lower and medium qualification requirements. There are also changes with regard to job characteristics. The number of part-time jobs and the demand for weekend and holiday work are reduced by climate protection policies.

Not being able to adequately capture innovation effects is one of the main weaknesses of modelling economic effects. Various theoretical paradigms such as neoclassical economics, evolutionary and institutional economics, and policy analysis highlight different mechanisms inducing technological change. Environmental economics argues within a rather linear model of sequential innovation stages: inventions lead to new technical developments which then diffuse through the market. There is a tendency to analyse the effects on the different stages of innovation separately. Assuming perfect economic rationality, decisions are based on microeconomic optimisation behaviour which is triggered by price changes. Evolutionary economics emphasises the generation of variety and selection. As far as certain developments have created favourable conditions for economic and technical change, an irreversible transition to new states then occurs through the use of temporary windows of opportunity. With regard to behaviour, the assumption of the perfect rationality of "homo economicus" is abandoned. Instead, behavioural routines which have evolved over a longer period of time play an important role, and replace the permanent optimisation due to smallest modifications in the frame conditons which dominates neoclassical theory. New institutional economics emphasises the fundamental significance of institutions for all aspects of economic behaviour. Policy analysis downplays the importance of microeconomic optimisation calculations, and implicitly uses a decision model, in which psychological elements seem to play an important part. Thus, it highlights the importance of "soft context factors" and downplays the importance of the instrument choice. It further follows that policies can have positive innovation effects, even if they use command-and-control instruments. In this respect, the policy analysis approach takes a counter position to orthodox environmental economics.

Findings from empirical analyses suggest that relative changes in environmentally-relevant costs influence the direction and the rate of innovation in environmental technologies. In addition, the case- and context-specific determinants for each individual subject of investigation do not allow the deduction of a general quantitative relationship for inducing environmental technology progress.

We now turn to the new empirical results presented in this book. Starting from the observation energy use in absolute and specific terms in the German manufacturing sector has dropped remarkably since the early 1970 we analyse the underlying determinants for this decline. To do so, total fuel intensity in the West German manufacturing sector was broken down into structural change effects (keeping subsector intensities constant) and efficiency effects (keeping industry structure constant) for the period 1970–1994. It could be shown that total energy intensity would have been much higher without structural change towards less energy-intensive outputs and without improved energy efficiency. *About three quarters of the observed reduction in total energy intensity can be attributed to energy-efficiency effects, while structural change effects account for about one quarter.*

A set of econometric analyses was then conducted to examine whether the observed decline in fuel intensity in the German manufacturing sector could be attributed to changes in fuel prices. First, energy intensity at total manufacturing and the structural change effect from the decomposition were regressed on various "explanatory" variables. The findings indicate that *higher fuel prices not only result in lower fuel intensity, but also trigger structural changes towards less energy-intensive products*. However, the magnitude of the price effect tends to be quite small. The results further suggest that higher investment rates not only reduce fuel intensity, but also trigger structural change towards less energy-intensive branches. Likewise, higher capacity use, which is assumed to reflect a booming economy, appears to increase the output share of more energy-intensive products. No well-specified equation was able to be made for the efficiency effect at the level of total manufacturing. A plausible explanation is that the impact of energy prices (and other factors) on energy intensity is likely to vary considerably across sub-sectors.

As a case study, the iron and steel sector in Germany was examined in more detail. First a regression equation for fuel intensity in pig iron production was estimated. The findings indicate that *higher fuel prices reduce fuel intensity, but the quantitative effect is rather small*, while autonomous technological change does not appear to have any impact at all. Production technologies in the iron and steel sector can be characterised as being of the "putty-clay" type. Thus, factor substitution in response to price changes is very limited, and companies determine their energy consumption when making choices from among different technologies for new investments. This would suggest that *the observed reduction in energy intensity is primarily the outcome of a switch in production technologies and of the diffusion of more energy-efficient technologies*.

Exploring econometrically the impact of various determinants on the diffusion of EAF-steel leads to the conclusion that material and energy input prices play a significant role in the decision to produce, using an alternative technological paradigm, the EAF technology. In addition, path dependencies stemming from high fixed costs dampen the diffusion of EAF-steel. The results of econometric analyses of the determinants for the observed fuel intensity and electricity intensity within the technological paradigms BOF-steel and EAF-steel suggest that higher energy and material input prices lead to lower energy intensity of the best-practice technologies. Furthermore, higher R&D expenditure by the mechanical and electrical engineering sectors and within the steel sector result in lower energy intensity. Finally, a higher degree of industry concentration dampens the adoption and diffusion of new energy-efficient technologies for the production of BOF-steel. To sum up, the results are generally consistent with the hypothesis of induced technological change. In terms of modelling, the findings suggest that induced technological change should be captured in energy-economic models estimating the costs of limiting greenhouse gas emissions. Otherwise, the costs of climate policies, such as higher energy taxes or emissions trading systems, will be overestimated by these models.[2]

In terms of policy implications, the findings suggest that modifying the tax scheme under the German Ecological Tax Reform to provide stronger incentives to save energy in the manufacturing sector would further reduce energy consumption.

[2] When using the DICE model of global warming, Popp (2004) finds that ignoring induced technological change in the case of a carbon tax overestimates the costs to society by almost 10%.

This would take place through a switch towards less energy-intensive products and production processes within and across sub-sectors, and through the accelerated adoption and diffusion of more energy-efficient technologies. In principle, similar effects may be expected from the EU-wide emissions trading system, which started in 2005 for most of the energy-intensive companies in the European Union. In an emissions trading system, the market price of a CO_2 allowance reflects the opportunity costs of increasing or reducing emissions. The higher the price of an allowance, the stronger is the incentive to reduce energy use. Allowance prices primarily depend on the stringency of the emission target that is on the number of allowances allocated and the reduction potential and costs of emission abatement measures.[3] However, specific allocation rules, such as the treatment of new entrants or plant closures or the rules on banking and borrowing of emission allowances across periods also have significant innovation effects.[4] While energy taxes and tradable emission allowance systems refer to the demand side of innovation, the estimation results also suggest that R&D in the technology supply sectors – and for BOF also in the steel sector – contributes to a better energy performance of new technologies.

The estimation results for the energy-intensive industry sectors are in line with the implications from innovation theory, in particular with the environmental economics literature. Companies from energy-intensive industries like the power, the iron and steel or the mineral processing industries tend to be quite aware of the potential cost savings from investing in energy efficiency. The high energy cost share in these companies provides a strong economic incentive to find and realise efficiency potentials. Likewise, since investing in energy efficiency directly affects the core production processes in energy-intensive companies, energy use is automatically considered in investment decisions. Such strong incentives however, do not exist in organisations where the energy cost share is relatively small. Techno-economic analyses suggest that there are tremendous energy savings opportunities in these organisations, but adoption of energy-efficient technologies has been slow, even where these technologies were cost-effective.

The empirical findings based on a large sample of organisations in the German commercial and services sectors support the view that energy consultations help to reduce barriers to the diffusion of energy efficient technologies and measures. The particular barriers considered are lack of time, lack of information about energy consumption patterns, lack of information about energy savings measures, organisational priority setting, uncertainty about energy costs and split incentives. More specifically, such consultations could be useful for SMEs or branches with low energy intensity, as these companies usually do not have recourse to their own energy experts and energy managers can only devote little time to

[3] See Betz et al. (2004, 2006) assessments of the allocation rules for the first and second phase of the EU Emissions Trading System (EU ETS).

[4] Schleich and Betz (2005) and Schleich et al. (2007) analyse the innovation effects of specific allocation rules across the EU for the first and second phase of the EU ETS, respectively.

energy efficiency (part time, at best). In this sense, the results rationalise support programmes for energy consultations as effective instruments to improve energy efficiency in these organisations.[5] In many countries, formal energy audits are subsidised via government or utility programmes that cover all or part of the costs for the audit. Subsidies are typically linked to the size of the organisation, such as the number of employees or annual turnover, energy consumption, energy costs, or particular sectors (WEC, 2001). Public support programmes tend to focus on small and medium-sized companies (SMEs). Some countries, including Portugal, Thailand and Tunisia, even require certain large industrial energy consumers to conduct regular energy audits. Some support programmes for investments in energy efficiency, such as the German ERP (European Recovery Programme) programme for Environment and Energy Saving, which provides low-interest loans to SMEs, require that the planned investment is recommended by an energy audit. In other countries, such as Denmark or Sweden, companies from the energy or industry sectors may be exempted from national energy taxes if they carry out an energy audit and implement the measures identified in these audits (Dyhr-Mikkelsen and Bach, 2005; Persson and Gudbjerg, 2005).

Interestingly, the findings also suggest that not all external energy efficiency consultants are equally effective. Engineering firms appear to be more successful than industrial sector associations or utilities. The most likely explanation is that information provided by industrial sector associations may be too general, and that utilities tend to focus on tariffs, rather than on technological or organisational measures to save energy costs. Results from former studies show only the limited success of an energy audit programme in Germany, which included small grants for energy audits in SMEs (Gruber and Venitz, 1994). Many small companies did not know the programme existed at all. Consequently, smaller companies were less likely to use the programme (Gruber and Brand, 1991). Others judged the grant to be too low because they were not able to assess in advance whether any future (uncertain) benefits of an audit would outweigh the costs. Most of them preferred a short but cost-free initial audit and wanted to pay the follow-up detailed audit on their own as soon as a reliable estimate about the saving potential existed.

Of course, energy audits alone cannot overcome all barriers. In the same way, the findings for the ENERGY variable indicate that price policies, such as taxes or emissions trading systems that raise the cost of energy use and increase the profitability of energy savings measures, are equally unlikely to suffice on their own.[6] Instead, as previous research has pointed out (Gruber and Brand, 1991; InterSEE, 1998; Sorrell et al., 2004), to accelerate the diffusion of energy efficiency a mix of well-targeted policy measures should be in place, by international, national or

[5] Since no data on costs and energy savings are available, it cannot be assessed whether the energy consultations were also efficient.

[6] Case study analyses also show that price policies and financial support programmes were only supporting factors among others and often not the most important one with regard to improving energy efficiency (InterSEE 1998).

regional policy makers, as well as by industrial sector associations, utilities, training organisations, research institutions and other groups that have a multiplier function. Increasingly, these measures also include new concepts for energy services including planning, implementation, financing and operating of energy-saving equipment (Chesshire, 2000; Schleich et al., 2001a; Sorrell, 2005).

In the fourth case study, the development of innovations in wind energy was analysed. The past 30 years have shown that reversals in the world leadership of wind power are possible to achieve. After an initial boom in the US in the 1970s, some European countries, notably Denmark, Germany and Spain, have now taken the lead. The analysis has shown that this can be attributed to an innovation system which successfully met the needs of the different phases of an evolving industry. Regulatory measures in the fields of R&D policies, environmental regulation and access to the monopolistic grid were responsible for this success. In addition to key functions such as market formation and supply of resources, the rapid market growth was also made possible by creating vicious cycles between the different functions of the innovation system which reinforced each other.

The analysis demonstrated that regulation is especially important in a case characterised by a triple regulation challenge in the three fields of spillovers of R&D, environmental protection, and access to monopolistic bottlenecks. Thus, especially for infrastructure sectors such as electricity, natural gas, transportation and possibly hydrogen in the future, the analysis of technological change cannot rely on single drivers such as energy prices, but must reflect the complex interplay of price changes, soft context factors and regulatory action. From a methodological point-of-view, it proved to be beneficial to use a sectoral/technological system of innovation as the heuristic framework for analysing these factors. The latest development in this string of research offers the possibility to link the role of regulation to innovation by analysing how the different functions of an innovation system are influenced. However, as the infrastructure sectors involved grabble with additional specifics such as long-term time horizon of investments and high path dependency of technological trajectories, it becomes a research challenge in itself to tailor the sectoral systems of innovation approach to the specifics of these sectors.

In general, the results confirm that there is no mechanistic relationship between policy and innovative effect. This indicates that a broad system concept is needed to analyse the innovation effects of environmental policy instruments. Thus, the concept of systems of innovation, which has been predominantly used to analyse national or technological systems, could also serve as a starting-point for analysing sustainability innovations as well. In particular, it may be necessary to move from the empirical analysis of complex innovation factors in case studies to results which can be generalised to a greater extent. Clearly, one of the key challenges for future research is evaluating the quantitative and statistical significance of "innovation-friendly soft context factors" and of specific innovation dynamics depending on systems of innovation behaviour which could then be incorporated into quantitative models to further explore the ramifications of the relationship between environmental policy and innovations.

It has been stated earlier in this summary that the innovation effects are not sufficiently considered in the macroeconomic model results. The detailed empirical analysis of the process of technological change in various energy related segments sheds light on the challenges ahead and possible ways for better integrating technological change in the macroeconomic models:

- For the energy intensive sectors, price induced technological change can explain large parts of technological change. At the same time, most macroeconomic models rely on price changes for all different forms of substitution effects. Thus, there is already a structural identity which allows for an expansion of the models with regard to integrating innovation without changing the basic logic of the models.
- For the non energy intensive sectors, the situation is more complex. The empirical results imply that soft context factors and various obstacles play a more important role. However, the new results also indicate that it is possible to account for these factors in econometric analysis. Thus, the challenge will be to use these approaches to forecast technological changes in these sectors. However, it is clear that, compared to the energy intensive sectors, this will result in approaches which expand the logic of the models towards factors not accounted for so far. Thus, an integration into the complex macroeconomic models will require much more additional research.
- The drivers and feedback mechanisms which prove to be important in the interplay of regulation and innovation are not accounted for at all in the approaches of the macroeconomic models. Thus, it won't be possible to integrate this kind of analysis in the macroeconomic models in the short run. Indeed, it will be a key challenge to construct an empirical bottom-up model which is able to explain the innovations in such circumstances. Perhaps alternative modelling approaches, such as systems dynamics or agent based modelling, which have been developed for areas other than economics, might offer a solution.[7]
- In addition to induced innovations, it will be necessary to include first mover advantages into the macroeconomic analysis. This task combines the challenges to account for induced innovations and the interplay between regulation and innovation. Furthermore, additional effects such as the importance of complementary industry clusters have to be accounted for. The next steps ahead call for more empirical work and bottom-up modelling of the dynamics of first mover advantages, before it will be possible to integrate these effects into macroeconomic models.

To sum up, the future research challenges in the field of macroeconomic effects of climate policy, structural adjustments and technological change imply a stepwise approach for the different sectors. In the short term, the energy intensive sectors are the ones which can be covered best within the complex macroeconomic models.

[7] System dynamic models have been hardly used in macroeconomics so far. One exception is the ASTRA-model (Schade 2005), which has been developed in conjunction with sector-specific models in the transportation sector.

For the non energy intensive sectors, an approach which builds on bottom-up analysis of technological change is feasible in the short run. The most difficult challenge will be the energy infrastructure sectors, and the complex interplay of regulation and innovation, especially in the context of first mover advantages. At least in the short run, combining the top down models with a bottom-up analysis, which is based on the insight of innovation research seems to be the most promising way of how to integrate innovation effects into macroeconomic analysis.

Appendix
Model Description of ISIS

ISIS (Integrated Sustainability Assessment System) is based on a static open multi-sectoral input-output-model and was developed at Fraunhofer ISI. It is used for integrated sustainability assessment of policy strategies and measures. In addition to the standard economic indicators such as economic output, value added, and employment, the model has been augmented with submodules for both socio-economic indicators (job structure, regional distribution of economic activities) and environmental pressure indicators (e.g. energy and resource use, greenhouse gases, other air emissions, wastewater and waste). A specific feature of the model is that it can be easily linked to systems and scenario analysis, using the results from these methods as input data for the model runs (impulse on final demand) on the one hand and information for the adaption of the model (introduction of new sectors, changes in the inter-industry matrices and of the various submodules) to case or technology specific circumstances on the other. Thus, the model is able to build a micro-macro bridge between a sound technological or systems analysis and the economic analysis.

More specifically, the model consists of the following building blocks:

- The *input-output model* is based on the input-output table for Germany, in which the economy is divided into 58 production sectors and 6 final demand sectors. The table incorporates the economic transactions between the production sectors, and between the production and the demand sectors. Thus, not only direct effects are taken into account, but also the indirect effects resulting from the inter-sectoral economic relations. Also, estimated technological progress resulting in productivity changes until 2020 is incorporated in the model. The input-output model is used to explore the sectoral differences in production between the reference scenario and the sustainability scenario. The model results form the basis for subsequent analyses.
- In the *module for quantitative employment effects*, employment coefficients are used to calculate sector-specific differences in employment between the reference scenario and the sustainability scenario. The model implicitly assumes that underemployment is the rule and that newly created positions can be filled without delay.
- In the *module for qualitative employment effects*, the results from the input-output model are linked with data from the German micro census survey to explore the

Sustainability and Innovation,

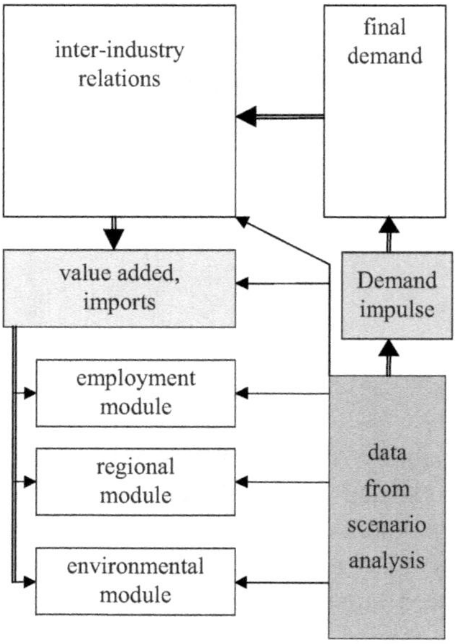

impact of increased car-sharing on qualification requirements, job characteristics, and working hours.

- In the *module for regional effects*, the results from the input-output model are linked with data from the German employment agency. This data set includes the sectoral distribution of employees for each of the 181 labour office districts. Based on this module, differences between the reference scenario and the sustainability scenario on regional concentration of the industry sectors can be analysed. In addition districts, which benefit and lose from the industrial ecology concepts, can be identified.

In the *module for environmental effects*, emission coefficients are used to calculate differences between the reference scenario and the sustainability scenario on primary energy consumption, the most important greenhouse gases, and air pollutants. Since indirect effects from inter-sectoral production relations are also taken into account, the modelling approach complies with the idea of a *life-cycle assessment (LCA)*. Furthermore, emission data for those processes responsible for the most important emissions are based on additional detailed analyses. Thus, in combining both the augmented input-output analysis and the LCA-type technology analysis, the advantages of both approaches are utilised (Duchin and Steenge, 1999).

Literature

Agterbosch S et al. (2004) Implementation of wind energy in the Netherlands: The importance of the social-institutional setting. Energy Policy 32: 2049–2066.

Aichinger H M, Mülheims K, Lüngen H B, Schierloh U, Stricker K P (2001) Ganzheitliche Bewertung und Potentiale der CO_2-Emission bei der Hochofen-Konverter-Route. Stahl und Eisen 121 (5): p 59–65.

Albrecht J (1998) Environmental regulation, comparative advantage, and the Porter Hypothesis. Nota di lavoro della Fondazio ENI Enrico Mattei Nr. 59. Mailand.

Albrecht J (2002) Environmental issue entrepreneurship: a Schumpeterian perspective, Futures 34 (7): 649–661.

Ameling D (2001) The importance of metallurgical coke for crude steel production. Stahl und Eisen 121 (11): p 31–37.

Ameling D, Aichinger H M (2001) Beitrag von Wirtschaft und Stahlindustrie zur Minderung klimawirksamer Emissionen in Deutschland im Kontext der Klimavorsorgepolitik. Stahl und Eisen 121 (7): p 61–70.

Ameling D (2004) Die Wettbewerbsfähigkeit der Stahlindustrie in Europa–Energiepolitik ein entscheidender Faktor, Pressegespräch mit der Wirtschaftspublizistischen Vereinigung am 28. Januar 2004 im Stahl-Zentrum Düsseldorf. http://www.stahl-online.de/medien_lounge/medieninformationen/Die_Wettbewerbsfhigkeit_der_Stahlindustrie_in_Europa.htm (Download October 5, 2004).

Amemiya T (1981) Qualitative response models: A survey. Journal of Economic Literature 19 (4): 483–536.

Arbeitsgemeinschaft Energiebilanzen (2003) Energy Balances. Preliminary results.

Archibugi D, Michie J (1998) Technical change, growth, and trade: New departures in institutional economics. Journal of Economic Surveys 12 (3): 313–332.

Arentsen M, Kemp R, Luiten E (2002) Technological change and innovation for climate protection: The challenge of governance. In: Kok M, Vermeulen W, Faaij A, de Jager D (eds.) Global warming and social innovation. London: Earthscan.

Arimura T et al. (2007) An empirical study of environmental R&D: What encourages facilities to be environmentally innovative? In: Johnstone N (ed.) Environmental policy and corporate behaviour. Edward Elgar, Cheltenham: 142–173.

Arrow K (1962) The economic implications of learning-by-doing, Review of Economic Studies 29 (6): 155–173.

Asheim B, Gertler M S (2005) The geography of innovation: Regional innovation systems. In: Fagerberg J et al. (eds.) The Oxford handbook of innovation. Oxford University Press, Oxford: 291–317.

Averch H, Johnson L (1962) Behavior of the firm under regulatory constraint. American Economic Review 52 (5): 1052–1069.

Barker T (1999) Achieving a 10% cut in Europe's carbon dioxide emissions using additional excise duties: Coordinated, uncoordinated and unilateral action using the econometric model E3ME. Economic Systems Research 11: 401–421.

Barker T, Johnstone N (1998) International competitiveness and carbon taxation. In: Barker T, Köhler J (eds.) International competitiveness and environmental policies. Edward Elgar, Cheltenham: 71–127.

Barro R (1990) Government spending in a simple model of endogenous growth. Journal of Political Economy 98: 103–125.

Baumol W J (1982) Contestable markets: An uprising in the theory of industry structure. American Economic Review 78: 1–15.

Beise M (2004) Lead markets: Country specific drivers of the global diffusion of innovations. Research Policy 33: 997–1028.

Beise M, Cleff T (2004) Assessing the lead market potential of countries for innovations projects. Journal of International Management 10 (4): 453–477.

Beise M, Rennings K (2003) Lead Markets of environmental innovations: A framework for innovation and environmental economics. ZEW Discussion Paper No. 03-01. Mannheim.

Beise M, Rennings K (2005) Lead markets and regulation: A framework for analyzing the international diffusion of environmental innovations. Ecological Economics 52 (1): 5–17.

Ben-David S, Brookshire D, Burness S, McKee M, Schmidt C (2000) Attitudes toward risk and compliance emission permit markets. Land Economics 76: 590–600.

Berg N, Holtbrügge D (1997) Wettbewerbsfähigkeit von Nationen: Der "Diamant"-Ansatz von Porter. WiSt 4: 199–201.

Bergek A, Jacobsson S (2003) The emergence of a growth industry: A comparative analysis of the German, Dutch and Swedish wind turbine industries. In: Metcalf S, Cantner U (eds.) Change, transformation and development. Physica-Verlag, Heidelberg: 197–227.

Betz R, Eichhammer W, Schleich J (2004) Designing national allocation plans for EU emissions trading – A first analysis of the outcomes. Energy & Environment 15: 375–425.

Betz R, Rogge K, Schleich J (2006) EU emission trading: An early analysis of national allocation plans for 2008–2012. Climate Policy 6: 361–394.

Bhattacharyya S C (1996) Applied general equilibrium models for energy studies: A survey. Energy Economics 18: 145–164.

Binswanger M (2001) Technological progress and sustainable development: What about the rebound effect? Ecological Economics 36: 119–132.

Bird L et al. (2005) Policies and market factors driving wind power development in the United States. Energy Policy 33: 1397–1407.

Blazejczak J (1987) Simulation gesamtwirtschaftlicher Perspektiven mit einem ökonometrischen Modell für die Bundesrepublik Deutschland. Beiträge zur Strukturforschung des DIW, Heft 100. Berlin.

Blazejczak J et al. (1999) Umweltpolitik und Innovation: Politikmuster und Innovationswirkungen im internationalen Vergleich, in: Zeitschrift für Umweltpolitik und Umweltrecht, Vol. 22, 1999, S. 1–32.

Blind K, Frietsch R (2005) Integration verschiedener Technologieindikatoren. Studien zum deutschen Innovationssystem Nr. 21-2005, BMBF. Berlin.

Blind K, Bührlen B, Menrad K, Hafner S, Walz R, Kotz C (2004) New products and services: Analysis of regulations shaping new markets. Office for Official Publications of the EU, Luxembourg.

Blümle G (1994) The importance of environmental policy for international competitiveness. In: Matsugi T, Oberhauser A (eds.) Interactions between economy and ecology. Berlin: Publisher is Duncker & Humblot: 35–57.

BMU (2004) Novelle des EEG am 1. August in Kraft getreten. Umwelt 2004 (9): 492–498.

BMWI (2000) Enegiedaten 2000. Bundesministerium für Wirtschaft, Bonn.

Böhringer C et al. (2001) Environmental taxation and structural change in an open economy. A CGE analysis with imperfect competition and free entry. ZEW Working Paper No. 01–07. Mannheim.

Bosello F et al. (2001) The double dividend issue: Modelling strategies and empirical findings. Environment and Development Economics 6: 9–45.

Bovenberg A L, de Mooij R A (1994) Environmental levies and distortionary taxation. American Economic Review 84: 1085–1089.

Bovenberg A L, van der Ploeg F (1994) Environmental policy, public finance and the labor market in a second-best world. Journal of Public Economics 55: 349–390.

Brauch H G (1996) Forschung, Entwicklung, Markteinführung und Exportförderung für erneuerbare Energien in den USA. In: Brauch H G (ed.) Energiepolitik. Springer, Heidelberg 1997, 15, 221–242.

Brechling V, Smith S (1994) Household energy efficiency in the UK. Fiscal Studies 15 (2): 44–56.

Brown M A (2001) Market failures and barriers as a basis for clean energy policies. Energy Policy 29: 1197–1207.

Brown M et al. (2000) Clean energy future for the US (5 National Laboratory Study). Study prepared by ORNL, LBNL, PNNL, NREL and ANL for the U.S. department of energy. Washington, DC.

Brunekreeft G (2004) Regulatory threat in vertically related markets: The case of German electricity. European Journal of Law and Economics 17: 285–305.

Buonanno P, Carraro C, Galeotti M (2003) Endogenous induced technical change and the costs of Kyoto. Resource and Energy Economics 25: 11–35.

Cameron L et al. (1999) The economics of strategies to reduce greenhouse gas emissions. Energy Studies Review 9: 63–73.

Cames M, Peter B, Seifried D, Lücking G, Matthes F (1996) Nachhaltige Energiewirtschaft – Einstieg in die Arbeitswelt von Morgen. Freiburg.

Carlsson B, Stankiewicz R (1995) On the nature, function and composition of technological systems. In: Carlsson B (ed.) Technological systems and economic performance: The case of factory automation. Kluwer Academic Publishers, Dordrecht.

Carlsson B, Jacobsson S, Holmen M, Rickne A (2002) Innovation systems: Analytical and methodological issues. Research Policy 31: 233–245.

Carraro C, Galeotti M (2002) Traditional environmental instruments, Kyoto mechanisms and the role of technical change. In: Carraro D, Egenhofer D (eds.) Firms, governments and climate policy – Incentive-based policies for long-term climate change. Edward Elgar, Cheltenham.

CEU (Council of the European Union) (2007, May 2) Presidency Conclusions. Brussels European Council 13–14 March 2007. Document 7224/1/07 REV 1. Brussels.

Chesshire J H (2000) From electrictiy supply to energy services. Prospects for Active Energy Services in the EU.

Coase R (1991) The nature of the firm (Reprint). In: Williamson O E, Winter S (eds.) The nature of the firm. Origins, evolution, and development. Oxford University Press, New York: 18–33.

Conrad K (1999) Computable general equilibrium models for environmental economics and policy analysis. In: van den Bergh J C J M (ed.) Handbook of environmental and resource economics. Edward Elgar, Cheltenham: 1060–1088.

Conrad K, Schmidt T (1999) Economic effects of an uncoordinated versus a coordinated carbon dioxide policy in the European Union: An applied general equilibrium analysis. Economic Systems Research 10: 161–182.

Conslik J (1996) Why bounded rationality? Journal of Economic Literature 34: 669–700.

DeAlmeida E L F (1998) Energy efficiency and the limits of market forces: The example of the electric motor market in France. Energy Policy 26: 643–653.

DeAlmeida A T, Fonseca P, Falkner H, Bertoldi P (2003) Market transformation of energy-efficient motor technologies in the EU. Energy Policy 31: 563–575.

DeCanio S J (1993) Barriers within firms to energy-efficient investments. Energy Policy 21: 906–914.

DeCanio S J (1994) Agency and control problems in US corporations: The case of energy efficient investment projects. Journal of the Economics of Business 1 (1): 105–123.

DeCanio S J (1998) The efficiency paradox: Bureaucratic and organisational barriers to profitable energy saving investments. Energy Policy 26: 441–454.
DeCanio S J (1999) Estimating the non-environmental consequences of greenhouse gas reduction is harder than you think. Contemporary Economic Policy 17: 279–295.
DeCanio S J, Watkins W E (1998a) Information processing and organizational structure. Journal of Economic Behavior and Organization 36: 275–294.
DeCanio S J, Watkins W E (1998b) Investment in energy efficiency: Do the characteristics of the firm matter? Review of Economics and Statistics 80: 95–107.
DeCanio S J et al. (2000) The importance of organizational structure for the adoption of innovations. Management Science 10: 1285–1299.
DeCanio S J et al. (2001) Organizational structure and the behavior of firms: Implications for integrated assessment. Climate Change 48: 487–514.
DeGroot H L F, Verhoef E T, Nijkamp P (2001) Energy savings by firms: Decision-making, barriers and policies. Energy Economics 23: 717–740.
Dennis M et al. (1990) Effective dissemination of energy-related information. American Psychologist 45: 1109–1117.
Destais G (1996) Economic effects of environmental policies and constraints: What can we learn from computable general equilibrium models? In: Faucheaux S et al. (eds.) Models of sustainable development. Edward Elgar, Cheltenham: 87–102.
DeVries F P, Withagen C (2005) Innovation and environmental stringency: The case of sulphur dioxide abatement. Discussion Paper No. 2005-18. University of Tilburg.
Diekmann J, Eichhammer W, Neubert A, Rieke H, Schlomann B, Ziesing H-J (1999) Energie-Effizienz-Indikatoren. Statistische Grundlagen, theoretische Fundierung und Orientierungsbasis für die politische Praxis. Physica, Heidelberg (short version available in English: http://www.eu.fhg.de/EEI/index.htm).
DIW (1994) Wirtschaftliche Auswirkungen einer ökologischen Steuerreform. Gutachten des DIW im Auftrag von Greenpeace. Berlin.
DIW et al. (2001) Die ökologische Steuerreform in Deutschland. Physica-Verlag, Heidelberg.
DIW/Fifo (1999) Anforderungen an und Anknüpfungspunkte für eine Reform des Steuersystems unter ökologischen Aspekten. Berichte des Umweltbundesamtes 3/99. Berlin.
DIW/Fifo/RWI/ZEW (1996) Der Einfluss von Energiesteuern und-abgaben zur Reduktion von Treibhausgasen auf Innovation und technischen Fortschritt. Clearing-Studie im Auftrag des BMBF. Berlin u a.
DIW/ISI/Berger R (2007) Wirtschaftsfaktor Umweltschutz: Vertiefende Analyse zu Umweltschutz und Innovation. Schriftenreihe Umwelt, Innovation, Beschäftigung des BMU/UBA, Nr. 01/07. Berlin.
DIW/ISI/Öko-Institut/WI/EWI/Prognos (2001) Energiewirtschaftliche Voraussetzungen und energiepolitische Handlungsmöglichkeiten für eine zukunftsfähige Energieentwicklung in Deutschland. Wissenschaftliche Begleitung des Energiedialogs 2000. Berlin.
Dixit A K, Pindyck R S (1994) Investment under uncertainty. Princeton University Press, Princeton, NJ.
Dosi G (1982) Technological paradigms and technological trajectories. A suggested interpretation of the determinants and directions of technical change. Research Policy 11: 147–162.
Dosi G (1988) The nature of the innovative process. In: Dosi G, Freeman C, Nelson R, Silverberg G, and Soete L (eds.) Technical change and economic theory. Pinter Publishers, London and New York: 221–238.
Dosi G, Pavitt K, Soete L (1990) The economics of technical change and international trade. Harvester Wheatsheaf, New York.
Duchin F, Steenge A E (1999) Input-output analysis, technology and the environment. In: van den Bergh J C J M (ed.) Handbook of environmental and resource economics. Edward Elgar, Cheltenham: 1037–1059.
Durstewitz M (2003, January) ISET. EXTOOL – EXCETP 6 – Workshop: Experience curves for wind and observed cost reductions. International Energy Agency. Paris [pdf].

Dyllick T, Hamschmidt J (1999) Wirkungen von Umweltmanagementsystemen – Eine Bestandsaufnahme empirischer Studien, Zeitschrift für Umweltpolitik und Umweltrecht: 507–540.

Dyhr-Mikkelsen K, Bach P (2005) Evaluation of free-of-charge energy audits. In: European council for an energy-efficient economy (Paris): Proceedings of the 2005 eceee Summer Study, 30 May–June 4. Mandelieu, France.

Edquist C (2005) Systems of innovation: Perspectives and challenges. In: Fagerberg J, Mowery D, Nelson R R (eds.) The Oxford handbook of innovation. Oxford University Press, Oxford: 181–208.

Edquist C, McKelvey M (eds.) (2000) Systems of innovation. Growth, competitiveness and employment. Elgar Reference Collection, Cheltenham.

Eggertsson (1990) Economic behaviour and institutions. Cambridge University Press, Cambridge.

Eichhammer W, Mannsbart W (2005) Business opportunities for producers of energy-efficient technologies and technologies using renewable energies. Report of Fraunhofer ISI, Karlsruhe.

Ekins P, Speck S (1998) The impacts of environmental policy on competitiveness: Theory and evidence. In: Barker T, Köhler J (eds.) International competitiveness and environmental policy. Edward Elgar, Cheltenham: 33–70.

Ender C (2002, August) International development of wind energy use – Status 31.12.2001. DEWI Magazin No. 21: 24–30.

Ender C (2004, August) International development of wind energy use – Status 31.12.2003. DEWI Magazin No. 25: 26–30.

Ender C (2005, August) International development of wind energy use – Status 31.12.2004. DEWI Magazin No. 27: 36–43.

Enquête Commission (1998) Protecting the earth's atmosphere. An international challenge. Interim Report of the Study Commission of the 11th German Bundestag "Preventive Measures to Protect the Earth's Atmosphere". Bundestag Bonn: Bonner Univ. Buchdruckerei.

Erdmann G (1993) Elemente einer evolutorischen Innovationstheorie. Tübingen.

European Commission (2001) The Guidebook for energy audits, programme schemes and administration procedures. SAVE-Project Final Report, DG TREN. Brussels.

European Commission (2006) Action plan for energy efficiency: Realising the potential. COM(2006)545 final.

EWAE (2004) Wind energy: The facts, European Wind Energy Association, Brussels.

Eyre N (1997) Barriers to energy efficiency: More than just market failure. Energy & Environment 8 (1): 25–43.

Fagerberg J (1994) Technology and international differences in growth rates. Journal of Economic Literature 32: 1147–1175.

Fagerberg J (1995a) Technology and competitiveness. Oxford Review of Economic Policy 12 (3): 39–51.

Fagerberg J (1995b) User-producer interaction, learning, and competitive advantage. Cambridge Journal of Economics 19: 243–256.

Fagerberg J, Godinho M M (2005) Innovation and catching-up. In: Fagerberg J et al. (eds.) The Oxford handbook of innovation. Oxford University Press, Oxford: 514–542.

Fankhauser S, McCoy D (1995) Modelling the economic consequences of environmental policies. In: Folmer H et al. (eds.) Principles of environmental and resource economics. Edward Elgar, Cheltenham: 253–275.

Faucheaux S, Levarlet F (1999) Energy–economy–environment models. In: van den Bergh J C J M (ed.) Handbook of environmental and resource economics. Edward Elgar, Cheltenham: 1123–1145.

Federal Environmental Agency (FEA) (Umweltbundesamt) (2002). Umweltdaten Deutschland 2002. Berlin.

Federal Ministry of Economics and Labour (FMEL) (Bundesministerium für Wirtschaft und Arbeit) (2002). Energiedaten 2002. Berlin.

Fischer C, Newell R (2004, April) Environmental and technology policies for climate change and renewable energy. RFF Discussion Paper No. 04–05, Resources for the Future. Washington, DC.
Foxon T J et al. (2005) UK innovation systems for new and renewable energy systems: Drivers, barriers and system failures. Energy Policy 33: 2123–2137.
Fri R W (2003) The role of knowledge: Technological innovation in the energy system. Energy Journal 24: 51–74.
Frohn J et al. (1998) Fünf makroökonometrische Modelle zur Erfassung der Wirkungen umweltpolitischer Maßnahmen – Eine vergleichende Betrachtung. Band 7 der Schriftenreihe "Beiträge zu den umweltökonomischen Gesamtrechnungen". Stuttgart.
Frohn J et al. (2003) Wirkungen umweltpolitischer Maßnahmen. Heidelberg.
Frondel M et al. (2007) What triggers environmental management and innovation? Empirical evidence for Germany. In: Ecological Economics, article in press, corrected proof, available 12 September 2007.
Garnreiter F et al. (1983) Auswirkungen verstärkter Maßnahmen zum rationellen Energieeinsatz auf Umwelt, Beschäftigung und Einkommen. Berichte des Umweltbundesamtes. Berlin.
Geiger B, Gruber E, Megele W (1999) Energieverbrauch und Energieeinsparung in Gewerbe, Handel und Dienstleistung. Physica, Heidelberg.
Gilchrist S, Williams J (2000) Putty-clay and investment: A business cycle analysis. Journal of Political Economy 108: 928–960.
Goulder L H (1995) Effects of carbon taxes in an economy with prior tax distortions: An intertemporal general equilibrium analysis. Journal of Environmental Economics and Management 29: 271–297.
Goulder L H, Matthai K (2000) Optimal CO_2 abatement in the presence of induced technological change. Journal of Environmental Economics and Management 39: 1–38.
Goulder L H, Schneider S H (1999) Induced technological change and the attractiveness of CO2-emission abatement. Resource and Energy Economics 21 (3–4): 211–253.
Grossman G, Helpman E (1990) Comparative advantage and long-run growth. American Economic Review 80: 796–815.
Grubb M et al. (1993) The costs of limiting fossil-fuel CO_2 emissions: A survey and analysis. Annual Review of Energy and the Environment 18: 397–478.
Gruber E, Brand M (1991) Promoting energy conservation in small and medium-sized companies. Energy Policy 19: 279–287.
Gruber E, Venitz J (1994) Energy conservation in small and medium-sized industry – Potentials, barriers, and policy instruments. Chapter 3 of the Project "Energy conservation – Opportunities not taken" for the commission of the European communities. Dublin and Karlsruhe.
Grupp H (1998) Foundations of the economics of innovation: Theory, measurement, and practice. Edward Elgar, Cheltenham.
Grupp H (1999) Umweltfreundliche Innovation durch Preissignale oder Regulation? Eine empirische Analyse für Deutschland. Jahrbücher für Nationalökonomie und Statistik 219: 611–631.
GWEC (2005, March) Global wind power continues expansion. Global Wind Energy Council. Brussels.
Haas R et al. (2004) How to promote renewable energy systems successfully and effectively. Energy Policy 32 (6): 833–839.
Hall B H (2004, January) Innovation and diffusion. NBER Working Paper No. 10212, National Bureau of Economic Research. Cambridge, MA.
Hall B H, Kahn B (2003) Adoption of new technology. UC Berkeley Department of Economics Working Paper No. E03–330. Berkeley, CA.
Hassett K A, Metcalf G E (1993) Energy conservation investment: Do consumers discount the future correctly? Energy Policy 21: 710–716.
Heal G (2008) Climate economics: A meta-review and some suggestions. National Bureau of Economic Research Working Paper No. 13927. NBER, Cambridge.
Hemmelskamp J (1998) Wind energy policy and their impact on innovation – An international comparison. Institute for Prospective Technology Studies. Seville, Spain.

Hemmelskamp J (1999) Innovationswirkungen der Umweltpolitik im Windenergiebereich. In: Klemmer P (ed.) Innovationen und Umwelt. Analytica, Berlin: 81–112.

Héritier A (ed.) (1993) Policy-analyse. Kritik und Neuorientierung. PVS-Sonderheft Nr. 24. Westdeutscher Verlag, Opladen.

Hicks J R (1932) The theory of wages. Macmillan, London.

Hippel E (1986) Lead users. A source of novel product concepts. Management Science 32: 791–805.

Hohmeyer O (2002) Vergleich externer Kosten der Stromerzeugung in Bezug auf das Erneuerbare Energien Gesetz. Texte des Umweltbundesamts Nr. 06/02. Berlin.

Hohmeyer O et al. (1985) Employment effects of energy conservation investments in EC countries. Fraunhofer ISI Karlsruhe / EG-Kommission, EG-Forschungsvorhaben EUR 109199 EN. Brüssel.

Horbach J (2004) Integrierte Technologien in neueren empirischen Untersuchungen. In: VDI (ed.) Umweltstatistiken – Einbeziehung von integrierten Technologien in Umweltstatistiken. Ergebnisse des Fachgesprächs am 14.11.2003 in Düsseldorf, Schriftenreihe des VDI Band 51. Düsseldorf.

Horbach J (ed.) (2005) Indicator systems for sustainable innovation. Springer Verlag, Heidelberg.

Horbach J (2007) Determinants of environmental innovation – New evidence from German panel data sources. Research Policy 37 (1): 163–173.

Hourcade J C, Robinson J (1996) Mitigating factors. Assessing the costs of reducing GHG emissions. Energy Policy 24: 863–873.

Howarth R B, Andersson B (1993) Market barriers to energy efficiency. Energy Economics 15: 262–272.

Howarth R B, Sanstad A H (1995) Discount rates and energy efficiency. Contemporary Economic Policy 13:101–109.

Howlett M, Ramesh M (1995) Studying public policy: Policy cycles and policy subsystems. Toronto/New York/Oxford.

InterSEE (1998) Interdisciplinary analysis of successful implementation of energy efficiency in industry, commerce and service. Wuppertal Institut für Klima Umwelt Energie, AKF-Institute for Local Government Studies, Energieverwertungsagentur, Fraunhofer ISI Karlsruhe, Projekt Klimaschutz am Institut für Psychologie der Universität Kiel, Amstein, Walthert, Energie B (eds.). Wuppertal, Kopenhagen, Wien, Karlsruhe, Kiel.

IPCC (1995) Second assessment report, WG III. Economic and social dimension of climate change. Cambridge University Press, Cambridge.

IPCC (2001) Third assessment report, WG III. Assessment of mitigation options. Cambridge University Press, Cambridge.

IPCC (Intergovernmental panel on climate change) (2001). Third assessment report. IPCC, Geneva.

IPCC (Intergovernmental panel on climate change) (2007). The physical science basis. Summary for policy makers. Contribution of Working Group I to the Fourth Assessment Report of the Intergovernmental Panel on Climate Change. IPCC, Geneva.

Jacobsson S, Johnson A (2000) The diffusion of renewable energy technology: An analytical framework and key issues for research. Energy Policy 28 (9): 625–640.

Jaffe A B, Stavins R N (1994a) Energy-efficiency investments and public policy. The Energy Journal 15 (2): 43–65.

Jaffe A B, Stavins R N (1994b) The energy-efficiency gap: What does it mean? Energy Policy 22: 804–810.

Jaffe A B et al. (1995) Environmental regulation and the competitiveness of U.S. manufacturing: What does the evidence tell us? Journal of Economic Literature 33: 132–163.

Jaffe A B, Palmer K (1997) Environmental Regulation and Innovation: A Panel Data Study, Review of Economics and Statistics, 79, p. 610–619.

Jaffe A B, Newell R G, Stavins R (2003) Technological change and the environment. In: Mäler K-G, Vincent J R (eds.) Handbook of environmental economics. Elsevier, Amsterdam: 461–516.

Jänicke M (ed.) (1996) Umweltpolitik der Industrieländer, edition sigma, Berlin.

Jänicke M, Weidner H (1995) Successful Environmental Policy. A Critical Evaluation of 24 Cases, edition sigma, Berlin.

Jänicke M et al. (1999) Innovationswirkungen branchenbezogener Regu-lierungsmuster am Beispiel energiesparender Kühlschränke in Dänemark. In: Klemmer P (ed.) Innovationen und Umwelt. Analytica, Berlin: 57–80.

Jochem E (1997a) Arbeit und bedachter Umgang mit Energie. In: Ropohl D, Schmid A (eds.) Handbuch zur Arbeitslehre. München: 687–699.

Jochem E (1997b) Some critical remarks on today's bottom-up energy models. In: Hake J-F, Markewitz P (eds.) Modellinstrumente für CO_2-Minderungs-strategien. Proceedings des Forschungszentrums. Jülich: 277–284.

Jochem E, Diekmann J (2001) Überlegungen zu einer sachgerechten Handhabung von Kostenangaben für Energiesystem-Modelle mit langfristigen Zeithorizonten. Arbeitspapier im Rahmen des IKARUS-Projektes, CEPE/ISI/DIW. Zürich/Karlsruhe/Berlin.

Jochem E, Eichhammer W (1999) Voluntary agreements as an instrument to substitute regulating and economic instruments. Lessons from the German voluntary agreements on CO2 reduction. In: Carraro C, Lévêque F (eds.) Voluntary approaches in environmental policy. Kluwer Academic Publisher, Dordrecht, Boston, London.

Jochem E, Gruber E (1990) Obstacles to rational electricity use and measures to alleviate them. Energy Policy 18: 340–350.

Jochem E, Schön M (1994) Gesellschaftliche und volkswirtschaftliche Auswirkungen der rationellen Energienutzung. Jahrbuch Arbeit und Technik. Bonn: 182–192.

Jochem E et al (1996) Exportchancen für Techniken zur Nutzung regenerativer Energien. Sachstandsbericht, TAB-Arbeitsberichte Nr. 42. Karlsruhe and Bonn.

Johnson A (1998) Functions in innovation system approaches. Mimeo, Department of Industrial Dynamics. Chalmers University of Technology, Sweden.

Jorgenson D W, Wilcoxen P J (1993) Energy prices, productivity, and economic growth. Annual Review of Energy 18: 343–395.

Kemp R (1997) Environmental Policy and Technical Change. A Comparison of the Technological Impact of Policy Instruments, Cheltenham, Edward Elgar.

Kemp R et al. (2000, May) How should we study the relationship between environmental regulation and innovation? IPTS Report EUR 19827 EN. Sevilla.

Kerr S, Newell R (2001, May) Policy-induced technology adoption: Evidence from the U.S. lead phasedown. RFF Discussion Paper No. 01-14, Resources for the Future. Washington DC.

Klemmer P, Lehr U, Löbbe K (1999) Umweltinnovationen, Analytica, Berlin.

Klemmer P (ed.) (1999) Innovation and the Environment, Analytica Verlag, Berlin.

Kline G J, Rosenberg N (1986) An overview of innovation. In: Landau R, Rosenberg N (eds.) The positive sum strategy: Harnessing technology for economic growth. Washington, DC: 275–306.

Knieps G (2001) The economics of network industries. In: Debreu G et al. (eds.) Economic essays. Springer, Berlin: 325–339.

Köhle S (1992) Einflussgrößen des elektrischen Energieverbrauchs und des Elektrodenverbrauchs von Lichtbogenöfen. Stahl und Eisen 112 (11): p 59–67

Köhle S (1999) Improvements in EAF operating practices over the last decade. Electric Furnace Conference Proceedings. Iron & Steel Society, Warrendale (1999).

Koomey J et al. (1998) Technology and GHG emissions: An integrated analysis using the LBNL-NEMS Model. Lawrence Berkeley National Laboratory Technical Report LBNL-42054. Berkeley, CA.

Koskela E, Schöb R, Sinn H-W (2001) Green tax reform and competitiveness. German Economic Review 2: 19–30.

Krause F (1996) The costs of mitigating carbon emissions: A review of methods and findings from European studies. Energy Policy 24: 899–915.

Krause F et al. (1999) Cutting carbon emissions while saving money: Low risk-strategies for the European Union. Volume II of the IPSEP Study Energy Policy in the Greenhouse, International Project for Sustainable Energy Paths. El Cerrito, CA.

Kuhlmann S, Arnold E (2001, November) RCN in the Norwegian research and innovation system. Reports in the evaluation of the research council of Norway. Fraunhofer ISI, Karlsruhe.
Laffont J-J, Tirole J (1986) Using cost information to regulate firms. Journal of Political Economy 94: 614–641.
Laitner S et al. (1998) Employment and other macroeconomic benefits of an innovation-led climate strategy for the United States. Energy Policy 26: 425–432.
Landjouw J O, Mody A (1996) Innovation and the International Diffusion of Environmentally Responsive Technology, in: Research Policy, Vol. 25, 1996, S. 549–571.
Landmann O (1984) Löhne, Preise, Einkommen, Beschäftigung in der offenen Volkswirtschaft. In: Bombach G et al. (eds.) Der Keynesianismus V. Springer, Heidelberg: 101–211.
Lauber V, Mez L (2004) Three decades of renewable energy politics in Germany. Energy & Environment 15 (4): 599–623.
Legler H et al. (1992) Innovationspotential und Hochtechnologie. Heidelberg.
Legler H, Schmoch U, Gehrke B, Krawczyk O (2002) Innovationsindikatoren zur Umweltwirtschaft, Studien zum deutschen Innovationssystem Nr. 2-2003. Niedersächsisches Institut für Wirtschaftsforschung. Hannover, Fraunhofer ISI, Karlsruhe.
Legler H, Krawczyk O, Walz R, Eichhammer W, Frietsch R (2006) Wirtschaftsfaktor Umwelt. Leistungsfähigkeit der deutschen Umwelt-und Klimaschutzwirtschaft im internationalen Vergleich. Texte des Umweltbundesamtes 16/06. Berlin.
Leone R, Hemmelskamp J (2000) The impact of EU regulation on innovation of European industry. Physica, Heidelberg.
Liaskas K, Mavrotas G, Mandaraka M, Diakoulaki D (2000) Decomposition of industrial CO_2 emissions: The case of the European Union. Energy Economics 22: 383–394.
Linder S B (1961) An essay on trade and transformation. Uppsala.
Linscheidt B, Truger A (2000) Ökologische Steuerreform: Ein Plädoyer für die Stärkung der Lenkungsanzreize. Wirtschaftsdienst (11): 98–106.
Lintz G (1992) Umweltpolitik und Beschäftigung. Beiträge zur Arbeitsmarkt-und Berufsforschung Band 159. Nürnberg.
Löschel A (2002) Technological change in economic models of environmental policy. A survey. Ecological Economics 43: 105–126.
Loulou R et al. (2000) Integration of GHG abatement options for Canada with the MARKAL model. Report prepared for the Canadian Government. Ottawa.
Lucas R E (1988) On the mechanics of economic development, Journal of Monetary Economics 22 (1): 3–42.
Lundvall B-A, Johnson B (1994) The learning economy. Journal of Industry Studies 1: 23–42.
Lundvall B-A et al. (2002) National systems of production, innovation and competence building. Research Policy 31: 213–231.
Lutz C, Meyer B, Nathani C, Schleich J (2005) Endogenous technological change and emissions: The case of the German steel industry. Energy Policy 33: 1143–1154.
Lutz C, Meyer B, Nathani C, Schleich J (2007) Endogenous innovation, economy and environment: Impacts of a technology-based modelling approach for energy-intensive industries in Germany. Energy Studies Review 15: 1–22.
Maddala G S (1983) Limited dependent and qualitative variables in econometrics. Cambridge University Press, Cambridge.
Maddala G S (1992) Introduction to econometrics, 2nd edition. Prentice Hall, Englewood Cliffs, NJ.
Malerba F (2002) Sectoral systems of innovation and production. Research Policy 31: 247–264.
Malerba F (2005) Sectoral systems – How and why innovation differs across sectors. In: Fagerberg J, Mowery D, Nelson R R (eds.) The Oxford handbook of innovation. Oxford University Press, Oxford: 380–406.
Masui T, Hanaoka T, Hikita S, Kainuma M (2006) Assessment of CO_2 Reductions and Economic Impacts Considering Energy-Saving Investments, The Energy Journal, Special Issue on Endogenous technological change and the economics of atmospheric stabilisation: 77–92.
Markard J et al. (2004) The impacts of market liberalization on innovation processes in the electricity sector. Energy & Environment 15 (2): 201–214.

McDonald R, Siegel D (1986) The value of waiting to invest. Quarterly Journal of Economics 101 (4): 707–727.
Metcalf G E (1994) Economics and rational conservation policy. Energy Policy 22: 819–825.
Meyer B, Ewerhart G (2001) INFORGE – Ein disaggregiertes Simulations-und Prognosemodell für Deutschland. In: Lorenz H-W, Meyer B (eds.) Studien zur evolutorischen Ökonomik IV. Duncker&Humblot, Berlin: 45–65.
Meyer B et al. (1997) Was kostet eine Reduktion der CO_2-Emissionen? Ergebnisse von Simulationsrechnungen mit dem umweltökonomischen Modell PANTA RHEI. Beiträge des Instituts für empirische Wirtschaftsforschung der Universität Osnabrück Nr. 55. Osnabrück.
Meyer B et al. (2001) Modelle und Ergebnisse – Das umweltökonomische Modell PANTA Rhei III. In: DIW et al. (eds.) Die ökologische Steuerreform in Deutschland. Physica, Heidelberg: 40–77.
Meyer B et al (2003) Simulationsergebnisse – PANTA Rhei: In: Frohn J et al. (eds.) Wirkungen umweltpolitischer Maßnahmen. Heidelberg: 121–147.
Meyer N I (2004) Development of Danish wind power market. Energy & Environment 15 (4): 657–673.
Michaelis P (1996) Ökonomische Instrumente der Umweltpolitik. Physica, Heidelberg.
Miketa A (2001) Analysis of energy intensity developments in manufacturing sectors in industrialized and developing countries. Energy Policy 29: 769–775.
Mitchell C, Connor P (2004) Renewable energy policy in the UK. Energy Policy 32: 1935–1947.
Montalvo C C (2002) Environmental policy and technological innovation. Edward Elgar, Cheltenham.
Morgan G (1985) Images of organisation. Sage, London.
Nelson R, Winter S (1982) An Evolutionary Theory of Economic Change. Harvard University Press. Cambridge (Massachusetts) and London.
Nelson R R (1994) The co-evolution of technology, industrial structure, and supporting institutions. Industrial and Corporate Change 3 (1): 47–63.
Nelson R R (1995) Recent evolutionary theorizing about economic change. Journal of Economic Literature 33: 48–90.
Nelson R R (2002) Technology, Institutions, and Innovation Systems. In: Research Policy 31: 265–272.
Newell R, Rogers K (2003, April) The market-based lead phasedown. RFF Discussion Paper No. 03-37, Resources for the Future. Washington, DC.
Newell R G, Jaffe A B, Stavins R N (1999) The induced innovation hypothesis and energy-saving technological change. Quarterly Journal of Economics 114: 941–975.
Nordhaus W D (2007) A review of the Stern Review on the economics of climate change. Journal of Economic Literature 45: 686–702.
OECD (2007) Renewables information 2007. International Energy Agency Statistics, Paris.
Ostertag K (2003) No regret potentials in energy conservation. Physica, Heidelberg.
Palmer K et al. (1995) Tightening environmental standards: The benefit-cost or the no-cost paradigm? Journal of Economic Perspectives 9: 119–132.
Panzar J C, Willig R D (1977) Free Entry and the Sustainability of Natural Monopoly, Bell Journal of Economics 8: pp 1–22.
Parry I W H, Bento A (2000) Tax-deductible spending, environmental policy, and the double dividend hypothesis. Journal of Environmental Economics and Management 39: 67–96.
Parry I W H, Williams R C, Goulder L H (1999) When can carbon abatement policies increase welfare? The fundamental role of distorted factor markets. Journal of Environmental Economics and Management 37: 52–84.
Pavitt K (1984) Sectoral patterns of technical change: Towards a taxonomy and a theory. Research Policy 13: 343–373.
Persson A, Gudjberg E (2005) Do voluntary agreements deliver? Experiences from Denmark and expectations for Sweden. In: European council for an energy-efficient economy (Paris). Proceedings of the 2005 eceee Summer Study, 30 May–June 4. Mandelieu, France.

Pfaffenberger W (1995) Arbeitsplatzeffekte von Energiesystemen. Frankfurt.

Pflüger M, Spermann A (1998) Ecological tax reform – A route to more employment. In: Schober F, Matsugi T (eds.) Labor market issues in Japan and Germany. Duncker&Humblot, Berlin: 101–121.

Popp D (2001) The effect of new technology on energy consumption. Resource and Energy Economics 23: 215–239.

Popp D (2002) Induced innovation and energy prices. American Economic Review 92: 160–180.

Popp D (2004) ENTICE: Endogenous technological change in the DICE model of global warming. Journal of Environmental Economics and Management 48: 742–768.

Porter M (1990) The competitive advantage of nations. New York.

Porter M, van der Linde C (1995) Toward a new conception of the environment-competitiveness relationship. Journal of Economic Perspectives 9: 97–118.

Posner M V (1961) International trade and technical change. Oxford Economic Papers 13: 323–341.

Prognos (2001) Klimaschutz und Arbeitsplätze. Peter Lang, Frankfurt.

Ragwitz M et al. (2005a, March) Analysis of the renewable energy sources' evolution up to 2020. Final Report of the FORRES Project to the CEC. Fraunhofer ISI, Karlsruhe.

Ragwitz M et al. (2005b, March) Energy scientific & technological indicators and references. Final Report of the ESTIR Project to the CEC. Fraunhofer ISI, Karlsruhe.

Ragwitz M et al. (2007) OPTRES. Assessment and optimisation of European renewable energy support schemes. IRB publishers, Stuttgart.

Ramesohl S (1998) Successful implementation of energy efficiency in light industry. In: International workshop on industrial energy efficiency policies: Understanding success and failure. Utrecht, 11–12 June, International network for energy demand analysis in the industrial sector (INEDIS). Proceedings Lawrence Berkeley National Laboratory (ed.) LBNL-42368.

Rehfeld K et al. (2007) Integrated product policy and environmental product innovations: An empirical analysis. Ecological Economics 61: 91–100.

Reiche D, Bechberger M (2004) Policy differences in the promotion of renewable energies in the EU member states. Energy Policy 32 (7): 843–849.

Rennings K et al. (2003) The influence of the EU environmental management and auditing scheme on environmental innovations and competitiveness in Germany. An analysis on the basis of a case studies and a large-scale survey. ZEW Discussion Papers No. Nr. 03-14. Mannheim.

Rennings K et al. (2006) The influence of different characteristics of the EU environmental management and auditing scheme on technical environmental innovations and economic performance. Ecological Economics 57: 45–59.

Reppelin-Hill V (1999) Trade and environment: An empirical analysis of the technology effect in the steel industry. Journal of Environmental Economics and Management 28: 283–301.

Richardson J J (1982) Policy styles in Western Europe. London.

Richter R (1994) Institutionen ökonomisch analysiert. Tübingen.

Richter R, Furubotn E G (1999) Neue Institutionenökonomik, 2nd edition. J. C. B. Mohr, Tübingen.

Romer P (1986) Increasing returns and long run growth. Journal of Political Economy 94: 1002–1037.

Romer P M (1990) Endogenous technological change. Journal of Political Economy 98: 71–102.

Ruth M, Amato A (2002) Vintage structure dynamics and climate change policies: The case of US iron and steel. Energy Policy 30: 541–552.

RWI/Ifo (1996) Gesamtwirtschaftliche Beurteilung von CO_2-Minderungsstrategien. Essen and München.

Sanstad A, Howarth R B (1994) Normal markets, market imperfections and energy efficiency. Energy Policy 22: 811–818.

Schade W (2005) Strategic sustainability analysis: Concept and application for the assessment of European transport policy. Nomos, Baden-Baden.

Schipper L et al. (2001) Carbon emissions from manufacturing energy use in 13 IEA countries: Long-term trends through 1995. Energy Policy 29: 667–688.

Schleich J (2001) The impact of fuel prices on energy intensity in the West German manufacturing sector, International Summer School on Economics, Innovation, Technological Progress and Environmental Policy, Kloster Seeon, September.

Schleich J (2004) Do energy audits help overcome barriers to energy efficiency? An empirical analysis for Germany. International Journal of Energy Technology and Policy: 226–239.

Schleich J (2007) Determinants of structural change and innovation in the German steel industry – An empirical investigation. International Journal of Public Policy 2: 109–123.

Schleich J, Betz R (2005) Incentives for energy efficiency and innovation in the EU emissions trading system. In: European council for an energy-efficient economy (Paris). Proceedings of the 2005 eceee Summer Study, 30 May–June 4. Mandelieu, France: 1459–1506.

Schleich J, Gruber E (2008) Beyond case studies: Barriers to energy efficiency in commerce and the services sectors. Energy Economics 30: 449–464.

Schleich J, Schlomann B, Eichhammer W (2000) Energy efficiency and structural change in the German manufacturing industry: In: European energy network: Monitoring tools for energy efficiency in Europe. Paris: ADEME Editions: 156–166.

Schleich J, Böde U, Köwener D, Radgen P (2001a) Chances and barriers for energy services companies – A comparative analysis for the German brewery and university sectors. In: European council for an energy-efficient economy (Paris). Proceedings of the 2001 eceee Summer Study. Further than ever from Kyoto? Rethinking energy efficiency can get us there. European council for an energy-efficient economy: 12: 229–240.

Schleich J, Eichhammer W, Böde U, Gagelmann F, Jochem E, Schlomann B, Ziesing H-J (2001b) Greenhouse gas reductions in Germany – Lucky strike or hard work. Climate Policy 1: 363–380.

Schleich J, Nathani C, Ostertag K, Schön M, Walz R, Meyer B, Lutz C, Distelkamp M, Hohmann F, Wolter M-I (2002) Innovationen und Luft schadstoffemissionen – Eine gesamtwirtschaftliche Abschätzung des Einflusses unterschiedlicher Rahmenbedingungen bei expliziter Modellierung der Technologiewahl im Industriesektor. Dokumentation Stahlindustrie. ISI/GWS, Karlsruhe, Osnabrück.

Schleich J, Nathani C, Meyer B, Lutz C (2006) Endogenous technological change and CO_2-emissions – The case of energy-intensive industries in Germany. Fraunhofer IRB Verlag, Stuttgart.

Schleich J, Betz R, Rogge K (2007) EU Emissions Trading – Better job second time around? In: European council for an energy-efficient economy (Paris). Proceedings of the 2007 eceee Summer Study, 4–9 June. La Colle sur Loup, France.

Schmidt A (1995) Ökonomische Auswirkungen rationeller Energieverwendung und erneuerbarer Energiequellen in den Bereichen Produktion und Außenhandel in Deutschland, 1976–1993. Arbeitspapiere ISI-A-2-95. FhG-ISI, Karlsruhe.

Schmidt T, Koschel H (1999) Beschäftigungswirkungen umweltpolitischer Instrumente zur Förderung integrierten Umweltschutzes. In: Pfeifer F, Rennings K (eds.) Beschäftigungswirkungen des Übergangs zu integrierter Umwelttechnik. Physica, Heidelberg: 153–172.

Schumpeter J A (1942) Capitalism, Socialism and Democracy (1942): New York: Harper & Row.

Schöb R (1995) Zur Bedeutung des Ökosteueraufkommens: Die Double-Dividend-Hypothese. Zeitschrift für Wirtschafts-und Sozialwissenschaften 115: 93–117.

Schön M, Ball M (2003) Eisen und Stahl, sector report for "Werkstoffeffizienz – Systemanalyse zu den Kreislaufpotenzialen energieintensiver Werkstoffe und ihrem Beitrag zur rationellen Energienutzung". Final Report for the Federal Ministry of Economics and Labour (BMWA = Bundesministerium für Wirtschaft und Arbeit). Fraunhofer ISI, Karlsruhe.

Scott S (1997) Household energy efficiency in Ireland: A replication study of ownership of energy saving items. Energy Economics 19: 187–208.

Seebregts A J et al. (2000) Endogenous technological change in energy systems models. International Journal of Global Energy Issues 14: 48–64.

Siebe T (1996) Sektorale Wirkungen der CO_2/Energiesteuer: Simulationen mit einem disaggregierten ökonometrischen Modell. Zeitschrift für Umweltpolitik und Umweltrecht 9: 454–468.
Silverberg G (1988) Modelling economic dynamics and technical change: Mathematical approaches to self-organisation and evolution. In: Dosi G, Freeman C, Nelson R, Silverberg G, Soete L (eds.) Technical change and economic theory. Pinter Publishers, London and New York.
Simon H A (1957) Models of man. Wiley, London.
Simon H A (1997) Models of bounded rationality, Vol. 3. Empirically grounded economic reason. MIT Press, Cambridge.
Smith A (2003) Transforming technological regimes for sustainable development: A role for alternative technology niches. Science & Public Policy 30 (2): 127–135.
Smits R, Kuhlmann S (2004) The rise of systemic instruments in innovation policy. International Journal of Foresight and Innovation Policy 1 (1): 1–26.
Smulders S (2001) Environmental taxation in open economies: In: Schulze G G, Ursprung H W (eds.) International environmental economics. Oxford: 166–182.
Sorrell S (2003) Making the link: Climate policy and the reform of the UK construction industry. Energy Policy 31: 865–878.
Sorrell S (2005) The economics of energy services contracting. In: European Council for an energy-efficient economy (Paris). Proceedings of the 2005 eceee Summer Study, 30 May–June 4. Mandelieu, France.
Sorrell S, O'Malley E, Schleich J, Scott S (2004) The economics of energy efficiency – Barriers to cost-effective investment. Edward Elgar, Cheltenham.
Spanos A (1990) Statistical foundations of econometric modelling. Cambridge University Press, Cambridge.
SRU Rat von Sachverständigen für Umweltfragen (2002) Umweltgutachten 2002 – für eine neue Vorreiterrolle, Metzler-Poeschel, Stuttgart.
Statistisches Bundesamt (2003) Statistisches Jahrbuch.
Stern N (2007) The economics of climate change. The Stern review. Cambridge University Press, Cambridge.
Stern P C (1986) Blind spots in policy analysis. What economics doesn't say about energy use. Journal of Policy Analysis and Management 5 (2): 200–227.
Stern P C (1992) What psychology knows about energy conservation. American Psychologist 47: 1224–1232.
Sun J W (1998) Changes in energy consumption and energy intensity: A complete decomposition model. Energy Economics 20: 85–100.
Taistra G (2001) Die Porter-Hypothese zur Umweltpolitik. Zeitschrift für Umweltpolitik und Umweltrecht: 241–262.
The Energy Journal (2006) Endogenous technological change and the economics of atmospheric stabilisation. Special Issue.
Thirtle C G, Rutan V W (1987) The Role of Demand and Supply in the Generation and Diffusion of Technical Change, Fundamentals of Pure and Applied Economics, Harwood Academic Publishers, New York.
Tol R S J (2006) The Stern review of the economics of climate change: A comment. Energy & Environment 17: 977–981.
Unander F et al. (1999) Manufacturing energy use in OECD countries: Decomposition of long-term trends. Energy Policy 27: 769–778.
UNDP/WEC/DESA (2000) World energy assessment. US department of energy (2005). Renewable portfolio standard overview, national renewable energy laboratory. Document DOE/GO 102005-2073, Golden, CO and Washington, DC. February 2005.
Utterback J M (1994) Mastering the dynamics of innovation: How companies can seize opportunities in the face of technological change. Harvard Business School Press: Boston, MA.
Vanberg V (2001) Rational choice vs. program-based behavior: Alternative theoretical approaches and their relevance for the study of institutions, Freiburger Diskussionspapiere zur Ordnungsökonomik, Nr. 01/05. Universität Freiburg, Freiburg.

Van der Zwaan B C C, Gerlagh R, Klaassen G, Schrattenholzer L (2002) Endogenous technological change in climate change modelling. Energy Economics 24: 1–19.

Van Soest D P, Bulte E H (2001) Does the energy-efficiency paradox exist? Technological progress and uncertainty. Environmental and Resource Economics 18: 101–112.

Vernon R (1966) International investment and international trade in the product cycle. Quarterly Review of Economics 88: 190–207.

Wakelin K (1997) Trade and innovation. Edward Elgar, Cheltenham.

Wagner M (2007) Empirical influence of environmental management on innovation: Evidence from Europe. In: Ecological economics, article in press, corrected proof, available 3 December 2007.

Wallace D (1995) Environmental Policy and Industrial Innovation, Earthscan, London.

Walz R (1994) Die Elektrizitätswirtschaft in den USA und der BRD. Physica-Verlag, Heidelberg.

Walz R (1995a) Gesamtwirtschaftliche Auswirkungen von Klimaschutzmaßnahmen – Der Modellierungsansatz der Enquete-Kommission. In: Hennicke P (ed.) Globale Kosten/Nutzen-Analysen von Klimaänderungen. Birkhäuser Verlag, Berlin: 134–152.

Walz R (1995b) Structural reforms in the electric utility industry: A comparison between Germany and the USA. ENER Bulletin 15: 53–71.

Walz R (1997) Economic impacts of climate protection policies: Results of a case study for Germany with a combined bottom-up/top down approach. In: Energy and sustainable growth: Is sustainable growth possible? Proceedings of the 20th International Association of Energy Economists International Conference, Vol. 1, New Delhi: 185–195.

Walz R (1999) Productivity effects of technology diffusion induced by an energy tax. Energy & Environment 10: 169–180.

Walz R, Dreher C, Marscheider-Weidemann F, Nathani C, Schirrmeister E, Schleich J, Schneider R, Schön M, Arbeitswelt in einer nachhaltigen Wirtschaft, UBA-Texte 44/01, Berlin.

Walz R (2002) Electric supply industry in Germany. In: De Paoli L (ed.) The electricity industry in transition. Franco Angeli Publications, Milano: 265–313.

Walz R (2007) The role of regulation for sustainable infrastructure innovations: The case of wind energy. International Journal of Public Policy 2: 57–88.

Walz R, Betz R (2003) Interaction of emissions trading with German climate policy instruments. Interaction in EU Climate Policy, Final Report. Fraunhofer ISI, Karlsruhe.

Walz R, Kotz C (2003, July) Innovation and regulation. Report within the EU-Project "Analysis of regulation shaping new markets". ISI-Report, Karlsruhe.

Walz R et al. (1995) Gesamtwirtschaftliche Auswirkungen von Emissionsminderungsstrategien. Studienprogramm der Enquête-Kommission "Schutz der Erdatmosphäre", Band 3 Energie. Teilband 2, Bonn.

Weber K M (1999) Innovation diffusion and political control of energy technologies – A comparison of combined heat and power generation in the UK and Germany. Physica, Heidelberg.

Welsch H (1996) Klimaschutz, Energiepolitik und Gesamtwirtschaft. Eine allgemeine Gleichgewichtsanalyse für die Europäische Union. München.

Welsch H (2001) The determinants of production-related carbon emissions in West Germany, 1985–1990. Assessing the role of technology and trade. University of Oldenburg, mimeo.

West G (1995) Comparison of input-output, input-output + econometric and computable general equilibrium models. Economic Systems Research 7: 209–227.

Williamson O E (1975) Markets and hierarchies: Some elementary considerations. Free Press, New York.

Williamson O E (1985) The economic institutions of capitalism. Free Press, New York.

Williamson O E, Winter S (1991) The nature of the firm – Origins, evolution, and development. New York and Oxford, Oxford University Press.

Wilson D, Swisher J (1993) Exploring the gap – Top-down versus bottom-up analyses of the cost of mitigating global warming. Energy Policy 21 (3): 249–263.

Wirtschaftsvereinigung Stahl and Verein Deutscher Eisenhüttenleute (WV Stahl and VDEH) Statistical yearbook of the steel industry different years. Düsseldorf, Stahleisen.

Witt U (1987) Individualistische Grundlagen der Evolutorischen Ökonomik. Tübingen.

Witt U (2003) The evolving economy - Essays on the evolutionary approach to economics, Edward Elgar, Aldershot.

World Energy Council (WEC) (2001) Energy efficiency policies and indicators. World Energy Council. London.

Xepapadeas A, de Zeeuw A (1999) Environmental policy and competitiveness: The Porter Hypothesis and the composition of capital. Journal of Environmental Economics and Management 37: 165–182.

Zajac E E (1970) A geometric treatment of Averch-Johnson's behavior of the firm model. American Economic Review 60: 117–125.

Ziesing H J, Jochem E, Mannsbart W, Schlomann B (1997) Causes of the trend in CO_2 emissions in Germany between 1990 to 1995. Executive summary. BMU, Bonn.

GPSR Compliance

The European Union's (EU) General Product Safety Regulation (GPSR) is a set of rules that requires consumer products to be safe and our obligations to ensure this.

If you have any concerns about our products, you can contact us on ProductSafety@springernature.com

In case Publisher is established outside the EU, the EU authorized representative is:

Springer Nature Customer Service Center GmbH
Europaplatz 3
69115 Heidelberg, Germany

Batch number: 10725973

Printed by Printforce, the Netherlands